D0226790

LEABHARLANN
CO. CILL DARA
910
4092

'An adopted national treasure after the publication
of *Notes from a Small Island* in which he pottered
around Britain, observing the country and its
characteristics with far more fondness and hilarity
than the way we Brits view ourselves'
Daily Telegraph

'Laugh-out-loud'
Observer

Withdrawn from
ATHY LIBRARY

'Not to be read in public, for fear of
emitting loud snorts'
The Times

'Seriously funny'
Sue Townsend, *Sunday Times*

'Hilarious'
Independent on Sunday

'Hugely funny (not snigger-snigger funny, but
great-big-belly-laugh-till-you-cry funny)'
Daily Telegraph

www.**billbryson**.co.uk

www.**transworldbooks**.co.uk

Bill Bryson's opening lines were:

'I come from Des Moines. Someone had to'.

This is what followed:

The Lost Continent

A road trip around the puzzle that is small-town America introduces the world to the adjective 'Brysonesque'.

> *'A very funny performance, littered with wonderful lines and memorable images'* LITERARY REVIEW

Neither Here Nor There

Europe never seemed funny until Bill Bryson looked at it.

> *'Hugely funny (not snigger-snigger funny but great-big-belly-laugh-till-you-cry funny)'* DAILY TELEGRAPH

Made in America

A compelling ride along the Route 66 of American language and popular culture gets beneath the skin of the country.

> *'A tremendous sassy work, full of zip, pizzazz and all those other great American qualities'* INDEPENDENT ON SUNDAY

Notes from a Small Island

A eulogy to Bryson's beloved Britain captures the very essence of the original 'green and pleasant land'.

> *'Not a book that should be read in public, for fear of emitting loud snorts'* THE TIMES

A Walk in the Woods

Bryson's punishing (by his standards) hike across the celebrated Appalachian Trail, the longest footpath in the world.

> *'This is a seriously funny book'* SUNDAY TIMES

Notes from a Big Country

Bryson brings his inimitable wit to bear on that strangest of phenomena – the American way of life.

> *'Not only hilarious but also insightful and informative'* INDEPENDENT ON SUNDAY

Down Under

An extraordinary journey to the heart of another big
country – Australia.

'*Bryson is the perfect travelling companion . . . When it comes to
travel's peculiars the man still has no peers*' THE TIMES

A Short History of Nearly Everything

Travels through time and space to explain the world, the
universe and everything.

'*Truly impressive . . . It's hard to imagine a better rough guide to
science*' GUARDIAN

The Life and Times of the Thunderbolt Kid

Quintessential Bryson – a funny, moving and perceptive
journey through his childhood.

'*He can capture the flavour of the past with the lightest of
touches*' SUNDAY TELEGRAPH

At Home

On a tour of his own house, Bill Bryson gives us an instructive
and entertaining history of the way we live.

'*A work of constant delight and discovery . . . don't leave home
without it*" SUNDAY TELEGRAPH

One Summer

Bryson travels back in time to a forgotten summer, when
America came of age, took centre stage and changed the world
for ever.

'*Has history ever been so enjoyable?*' MAIL ON SUNDAY

The Road to Little Dribbling

Two decades after *Notes from a Small Island*, Bill Bryson takes a
new amble round Britain, to rediscover the beautiful, eccentric
and endearing country he calls home.

'*Is it the funniest travel book I've read all year? Of course it is*'
DAILY TELEGRAPH

By

Bill Bryson

The Lost Continent

Mother Tongue

Troublesome Words

Neither Here Nor There

Made in America

Notes from a Small Island (published in the
USA as *I'm a Stranger Here Myself*)

A Walk in the Woods

Notes from a Big Country

Down Under (published in the USA as *In a
Sunburned Country*)

African Diary

A Short History of Nearly Everything

The Life and Times of the Thunderbolt Kid

Shakespeare (Eminent Lives Series)

Bryson's Dictionary for Writers and Editors

Icons of England

At Home

One Summer

The Road to Little Dribbling

BILL BRYSON

The Life and Times of the
THUNDERBOLT
KID

TRAVELS THROUGH MY CHILDHOOD

BLACK SWAN

TRANSWORLD PUBLISHERS
61–63 Uxbridge Road, London W5 5SA
www.transworldbooks.co.uk

Transworld is part of the Penguin Random House group of companies
whose addresses can be found at global.penguinrandomhouse.com

First published in Great Britain in 2003 by Doubleday
an imprint of Transworld Publishers
Black Swan edition published 2004
Black Swan edition reissued 2015

Copyright © Bill Bryson 2003
Illustrations by Neil Gower

Bill Bryson has asserted his right under the Copyright,
Designs and Patents Act 1988 to be identified as the author of this work.

Every effort has been made to obtain the necessary permissions with
reference to copyright material, both illustrative and quoted. We apologize
for any omissions in this respect and will be pleased to make the appropriate
acknowledgements in any future edition.

A CIP catalogue record for this book
is available from the British Library.

ISBN
9781784161811

Typeset in 11/16.25pt Giovanni by Falcon Oast Graphic Art Ltd.

Penguin Random House is committed to a sustainable future for
our business, our readers and our planet. This book is made from
Forest Stewardship Council® certified paper.

MIX
Paper from
responsible sources
FSC® C018179

Printed and bound in Great Britain by Clays Ltd, St Ives plc

3 5 7 9 10 8 6 4

Contents

Preface and Acknowledgements

My kid days were pretty good ones, on the whole. My parents were patient and kind and approximately normal. They didn't chain me in the cellar. They didn't call me 'It'. I was born a boy and allowed to stay that way. My mother, as you'll see, sent me to school once in Capri pants, but otherwise there was little trauma in my upbringing.

Growing up was easy. It required no thought or effort on my part. It was going to happen anyway. So what follows isn't terribly eventful, I'm afraid. And yet it was by a very large margin the most fearful, thrilling, interesting, instructive, eye-popping, lustful, eager, troubled, untroubled, confused, serene and unnerving time of my life. Coincidentally, it was all those things for America, too.

Everything recorded here is true and really happened, more or less, but nearly all the names and a few of the details have been changed in the hope of sparing embarrassment. A small part of the story originally appeared in somewhat different form in the *New Yorker*.

As ever, I have received generous help from many quarters, and I would like to thank here, sincerely and alphabetically, Deborah Adams, Aosaf Afzal, Matthew Angerer, Charles Elliott, Larry Finlay, Will Francis, Carol Heaton, Jay Horning, Patrick Janson-Smith, Tom and

Nancy Jones, Sheila Lee, Fred Morris, Steve Rubin, Marianne Velmans, Daniel Wiles, and the staff of the Drake University and Des Moines Public Libraries in Iowa and Durham University Library in England.

I remain especially grateful to Gerry Howard, my astute and ever thoughtful American publisher, for a stack of Boys' Life magazines, one of the best and most useful gifts I have had in years, and to Jack Peverill of Sarasota, Florida, for the provision of copious amounts of helpful material. And of course I remain perpetually grateful to my family, not least my dear wife, Cynthia, for more help than I could begin to list, to my brother Michael, and to my incomparably wonderful, infinitely sporting mother, Mary McGuire Bryson, without whom, it goes without saying, nothing that follows would have been possible.

Chapter 1
HOMETOWN

SPRINGFIELD, ILL. (AP) – The State Senate of Illinois yesterday disbanded its Committee on Efficiency and Economy 'for reasons of efficiency and economy'.

– *Des Moines Tribune*, 6 February 1955

IN THE LATE 1950s, the Royal Canadian Air Force produced a booklet on isometrics, a form of exercise that enjoyed a short but devoted vogue with my father. The idea of isometrics was that you used any unyielding object, like a tree or wall, and pressed against it with all your might from various positions to tone and strengthen different groups of muscles. Since everybody already has access to trees and walls, you didn't need to invest in a lot of costly equipment, which I expect was what attracted my dad.

What made it unfortunate in my father's case was that he would do his isometrics on aeroplanes. At some point in every flight, he would stroll back to the galley area or the space by the emergency exit and, taking up the posture of someone trying to budge a very heavy piece of machinery, he would begin to push with his back or shoulder against the outer wall of the plane, pausing occasionally to take deep breaths before returning with quiet, determined grunts to the task.

Since it looked uncannily, if unfathomably, as if he were trying to force a hole in the side of the plane, this naturally drew attention. Businessmen in nearby seats would stare over the tops of their glasses. A stewardess would pop her head out of the galley and likewise stare, but with a certain hard caution, as if remembering some aspect of her training that she had not previously been called upon to implement.

Seeing that he had observers, my father would straighten up, smile genially and begin to outline the engaging principles behind isometrics. Then he would give a demonstration to an audience that swiftly consisted of no one. He seemed curiously incapable of feeling embarrassment in such situations, but that was all right because I felt enough for both of us – indeed, enough for us and all the other passengers, the airline and its employees, and the whole of whatever state we were flying over.

Two things made these undertakings tolerable. The first was that back on solid ground my dad wasn't half as foolish most of the time. The second was that the purpose of these trips was always to go to a big city like Detroit or St Louis, stay in a large hotel and attend ballgames, and that excused a great deal – well, everything, in fact. My dad was a sportswriter for the *Des Moines Register*, which in those days was one of the country's best papers, and often took me along on trips through the Midwest. Sometimes these were car trips to smaller places like Sioux City or Burlington, but at least once a summer we boarded a silvery plane – a huge event in those days – and lumbered through the summery skies, up among the fleecy clouds, to a proper metropolis to watch Major League baseball, the pinnacle of the sport.

Like everything else in those days, baseball was part of a simpler world, and I was allowed to go with him into the changing rooms and dugout and on to the field before games. I have had my hair tousled by Stan Musial. I have handed Willie Mays a ball that had skittered past him as he played catch. I have lent my binoculars to Harvey Kuenn (or possibly it was Billy Hoeft) so that he could scope some busty blonde in the upper deck. Once on a hot July afternoon I sat in a nearly airless clubhouse under the left field grandstand at Wrigley Field in Chicago beside Ernie Banks, the Cubs' great shortstop, as he autographed boxes of new white baseballs (which are, incidentally, the most pleasurably aromatic things on earth, and worth spending time around anyway). Unbidden, I took it upon myself to sit beside him and pass him each new ball. This slowed the process considerably, but he gave a little smile each time and said thank you as if I had done him quite a favour. He was the nicest human being I have ever met. It was like being friends with God.

I can't imagine there has ever been a more gratifying time or place to be alive than America in the 1950s. No country had ever known such prosperity. When the war ended the United States had $26 billion worth of factories that hadn't existed before the war, $140 billion in savings and war bonds just waiting to be spent, no bomb damage and practically no competition. All that American companies had to do was stop making tanks and battleships and start making Buicks and Frigidaires – and boy did they. By 1951, when I came sliding down the chute, almost 90 per cent of American families had refrigerators, and nearly three quarters had washing machines, telephones, vacuum cleaners and gas or electric stoves – things that most of the rest of the world could still only fantasize about. Americans owned 80 per cent of the world's electrical goods, controlled two-thirds of the world's productive capacity, produced over 40 per cent of its electricity, 60 per cent of its oil and 66 per cent of its steel. The 5 per cent of people on Earth who were Americans had more wealth than the other 95 per cent combined.

I don't know of anything that better conveys the happy bounty of the age than a photograph (reproduced in this volume on pages 4 and 5) that ran in *Life* magazine two weeks before my birth. It shows the Czekalinski family of Cleveland, Ohio – Steve, Stephanie and two sons, Stephen and Henry – surrounded by the two and a half tons of food that a typical blue-collar family ate in a year. Among the items they were shown with were 450 pounds of flour, 72 pounds of shortening, 56 pounds of butter, 31 chickens, 300 pounds of beef, 25 pounds of carp, 144 pounds of ham, 39 pounds of coffee, 690 pounds of potatoes, 698 quarts of milk, 131 dozen eggs, 180 loaves of bread, and 8½ gallons of ice cream, all purchased on a budget of $25 a week. (Mr Czekalinski made $1.96 an hour as a shipping clerk in a Du Pont factory.) In 1951, the average American ate 50 per cent more than the average European.

No wonder people were happy. Suddenly they were able to have things they had never dreamed of having, and they couldn't believe their luck. There was, too, a wonderful simplicity of desire. It was the last time that people would be thrilled to own a toaster or waffle iron. If you bought a major appliance, you invited the neighbours round to have a look at it. When I was about four my parents bought an Amana Stor-Mor refrigerator and for at least six months it was like an honoured guest in our kitchen. I'm sure they'd have drawn it up to the table at dinner if it hadn't been so heavy. When visitors dropped by unexpectedly, my father would say: 'Oh, Mary, is there any iced tea in the Amana?' Then to the guests he'd add significantly: 'There usually is. It's a Stor-Mor.'

'Oh, a Stor-Mor,' the male visitor would say and raise his eyebrows in the manner of someone who appreciates quality cooling. 'We thought about getting a Stor-Mor ourselves, but in the end we went for a Philco Shur-Kool. Alice loved the E-Z Glide vegetable drawer and you can get a full quart of ice cream in the freezer box. *That* was a big selling point for Wendell Junior, as you can imagine!'

They'd all have a good laugh at that and then sit around drinking iced tea and talking appliances for an hour or so. No human beings had ever been quite this happy before.

People looked forward to the future, too, in ways they never would again. Soon, according to every magazine, we were going to have underwater cities off every coast, space colonies inside giant spheres of glass, atomic trains and airliners, personal jetpacks, a gyrocopter in every driveway, cars that turned into boats or even submarines, moving sidewalks to whisk us effortlessly to schools and offices, dome-roofed automobiles that drove themselves along sleek superhighways allowing Mom, Dad and the two boys (Chip and Bud or Skip and Scooter) to play a board game or wave to a neighbour in a passing gyrocopter or just sit back and enjoy saying some of those delightful words that existed in the Fifties and are no longer heard: *mimeograph, rotisserie, stenographer, ice box, rutabaga, panty raid, bobby sox, sputnik, beatnik, canasta, Cinerama, Moose Lodge, pinochle, daddy-o.*

For those who couldn't wait for underwater cities and self-driving cars, thousands of smaller enrichments were available right now. If you were to avail yourself of all that was on offer from advertisers in a single issue of, let's say, *Popular Science* magazine from, let's say, December 1956, you could, among much else, teach yourself ventriloquism, learn to cut meat (by correspondence or in person at the National School of Meat Cutting in Toledo, Ohio), embark on a lucrative career sharpening skates door to door, arrange to sell fire extinguishers from home, end rupture troubles once and for all, build radios, repair radios, perform on radio, talk on radio to people in different countries and possibly different planets, improve your personality, get a personality, acquire a manly physique, learn to dance, create personalized stationery for profit, or 'make $$$$' in your spare time at home building lawn figures and other novelty ornaments.

My brother, who was normally quite an intelligent human being, once invested in a booklet that promised to teach him how to throw his voice. He would say something unintelligible through rigid lips, then quickly step aside and say, 'That sounded like it came from over there, didn't it?' He also saw an ad in *Mechanics Illustrated* that invited him to enjoy colour television at home for 65 cents plus postage, placed an order and four weeks later received in the mail a multi-coloured sheet of transparent plastic that he was instructed to tape over the television screen and watch the image through.

Having spent the money, my brother refused to concede that it was a touch disappointing. When a human face moved into the pinkish part of the screen or a section of lawn briefly coincided with the green portion, he would leap up in triumph. 'Look! Look! *That's* what colour television's gonna look like,' he would say. 'This is all just experimental, you see.'

In fact, colour television didn't come to our neighbourhood until nearly the end of the decade, when Mr Kiessler on St John's Road bought an enormous RCA Victor Consolette, the flagship of the RCA fleet, for a lot of money. For at least two years his was the only known colour television in private ownership, which made it a fantastic novelty. On Saturday evenings the children of the neighbourhood would steal into his yard and stand in his flowerbeds to watch a programme called *My Living Doll* through the double window behind his sofa. I am pretty certain that Mr Kiessler didn't realize that two dozen children of various ages and sizes were silently watching the TV with him or he wouldn't have played with himself quite so enthusiastically every time Julie Newmar bounded on to the screen. I assumed it was some sort of isometrics.

Every year for nearly forty years, from 1945 until his retire-ment, my father went to the baseball World Series for the *Register*. It was, by an immeasurably wide margin, the high point of his working year. Not only did he get to live it up for two weeks on expenses in some of the nation's most cosmopolitan and exciting cities – and from Des Moines all cities are cosmopolitan and exciting – but he also got to witness many of the most memorable moments of baseball history: Al Gionfriddo's miraculous one-handed catch of a Joe DiMaggio line drive, Don Larsen's perfect game in 1956, Bill Mazeroski's series-winning home run of 1960. These will mean nothing to you, I know – they would mean nothing to most people these days – but they were moments of near ecstasy that were shared by a nation.

In those days, World Series games were played during

the day, so you had to bunk off school or develop a convenient chest infection ('Jeez, Mom, the teacher said there's a lot of TB going around') if you wanted to see a game. Crowds would lingeringly gather wherever a radio was on or a TV played. Getting to watch or listen to any part of a World Series game, even half an inning at lunchtime, became a kind of illicit thrill. And if you did happen to be there when something monumental occurred, you would remember it for the rest of your life. My father had an uncanny knack for being present at such moments – never more so than in the seminal (and what an apt word that can sometimes be) season of 1951 when our story begins.

In the National League (one of two principal divisions in Major League baseball, the other being the American League) the Brooklyn Dodgers had been cruising towards an easy championship when, in mid-August, their crosstown rivals the New York Giants stirred to life and began a highly improbable comeback. Suddenly the Giants could do no wrong. They won thirty-seven of forty-four games down the home stretch, cutting away at the Dodgers' once-unassailable lead in what began to seem a fateful manner. By mid-September people talked of little else but whether the Dodgers could hold on. Many dropped dead from the heat and excitement. The two teams finished the season in a perfect dead heat, so a three-game playoff series was hastily arranged to determine who would face the American League champions in the World Series. The *Register*, like nearly all distant papers, didn't dispatch a reporter to these impromptu playoffs, but elected to rely on wire services for its coverage until the Series proper got under way.

The playoffs added three days to the nation's exquisite torment. The two teams split the first two games, so it came down to a third, deciding game. At last the Dodgers appeared to recover their former poise and invincibility. They took a comfortable 4–1 lead into the final inning, and needed just three outs to win. But the Giants struck back, scoring a run and putting two more runners on base when Bobby Thomson (born in Glasgow, you may be proud to know) stepped to the plate. What Thomson did that afternoon in the gathering dusk of autumn has been many times voted the greatest moment in baseball history.

'Dodger reliever Ralph Branca threw a pitch that made history yesterday,' one of those present wrote. 'Unfortunately it made history for someone else. Bobby Thomson, the "Flying Scotsman," swatted Branca's second offering over the left field wall for a game-winning home run so momentous, so startling, that it was greeted with a moment's stunned silence.

'Then, when realization of the miracle came, the double-decked stands of the Polo Grounds rocked on their 40-year-old foundations. The Giants had won the pennant, completing one of the unlikeliest comebacks baseball has ever seen.'

The author of those words was my father – who was abruptly, unexpectedly, present for Thomson's moment of majesty. Goodness knows how he had talked the notoriously frugal management of the *Register* into sending him the one thousand one hundred and thirty-two miles from Des Moines to New York for the crucial deciding game – an act of rash expenditure radically out of keeping with decades of careful precedent – or how he had managed to secure credentials and a place in the press box at such a late hour.

But then he had to be there. It was part of his fate, too. I am not *exactly* suggesting that Bobby Thomson hit that home run because my father was there or that he wouldn't have hit it if my father had not been there. All I am saying is that my father was there and Bobby Thomson was there and the home run was hit and these things couldn't have been otherwise.

My father stayed on for the World Series, in which the Yankees beat the Giants fairly easily in six games – there was only so much excitement the world could muster, or take, in a single autumn, I guess – then returned to his usual quiet life in Des Moines. Just over a month later, on a cold, snowy day in early December, his wife went into Mercy Hospital and with very little fuss gave birth to a baby boy: their third child, second son, first superhero. They named him William, after his father. They would call him Billy until he was old enough to ask them not to.

* * *

Apart from baseball's greatest home run and the birth of the Thunderbolt Kid, 1951 was not a hugely eventful year in America. Harry Truman was President, but would shortly make way for Dwight D. Eisenhower. The war in Korea was in full swing and not going well. Julius and Ethel Rosenberg had just been notoriously convicted of spying for the Soviet Union, but would sit in prison for two years more before being taken to the electric chair. In Topeka, Kansas, a mild-mannered black man named Oliver Brown sued the local school board for requiring his daughter to travel twenty-one blocks to an all-black school when a perfectly good white one was just seven blocks away. The case, immortalized as *Brown v. the Board of Education*, would be one of the most far-reaching in modern American history, but wouldn't become known outside jurisprudence circles for another three years when it reached the Supreme Court.

America in 1951 had a population of one hundred and fifty million, slightly more than half as much as today, and only about a quarter as many cars. Men wore hats and ties almost everywhere they went. Women prepared every meal more or less from scratch. Milk came in bottles. The postman came on foot. Total government spending was $50 billion a year, compared with $2,500 billion now.

I Love Lucy made its television debut on 15 October, and Roy Rogers, the singing cowboy, followed in December. In Oak Ridge, Tennessee, that autumn police seized a youth on suspicion of possessing narcotics when he was found with some peculiar brown powder, but he was released when it was shown that it was a new product called instant coffee. Also new, or not quite yet invented, were ball-point pens, fast foods, TV dinners, electric can openers, shopping malls, freeways, supermarkets, suburban sprawl, domestic air conditioning, power steering, automatic transmissions, contact lenses, credit cards, tape recorders, garbage disposals, dishwashers, long-playing records, portable record players, Major League baseball teams west of St Louis, and the hydrogen bomb. Microwave ovens were available, but weighed seven hundred pounds. Jet travel, Velcro, transistor radios and computers smaller than a small building were all still some years off.

Nuclear war was much on people's minds. In New York on Wednesday 5 December, the streets became eerily empty for seven minutes as the city underwent 'the biggest air raid drill of the atomic age', according to *Life* magazine, when a thousand sirens blared and people scrambled (well, actually walked jovially, pausing upon request to pose for photographs) to designated shelters, which meant essentially the inside of any reasonably solid building. *Life*'s photos showed Santa Claus happily leading a group of children out of Macy's, half-lathered men and their barbers trooping out of barber shops, and curvy models from a swimwear shoot shivering and feigning good-natured dismay as they emerged from their studio, secure in the knowledge that a picture in *Life* would do their careers no harm at all. Only restaurant patrons were excused from taking part in the exercise on the grounds that New Yorkers sent from a restaurant without paying were unlikely to be seen again.

Closer to home, in the biggest raid of its type ever undertaken in Des Moines, police arrested nine women for prostitution at the old Cargill Hotel at Seventh and Grand downtown. It was quite an operation. Eighty officers stormed the building just after midnight, but the hotel's resident ladies were nowhere to be found. Only by taking exacting measurements were the police able to discover, after six hours of searching, a cavity behind an upstairs wall. There they found nine goose-pimpled, mostly naked women. All were arrested for prostitution and fined $1,000 each. I can't help wondering if the police would have persevered quite so diligently if it had been naked men they were looking for.

The eighth of December 1951 marked the tenth anniversary of America's entry into the Second World War, and the tenth anniversary plus one day of the Japanese attack on Pearl Harbor. In central Iowa, it was a cold day with light snow and a high temperature of 28°F/−2°C but with the swollen clouds of a blizzard approaching from the west. Des Moines, a city of two hundred thousand people, gained ten new citizens that day – seven boys and three girls – and lost just two to death.

Christmas was in the air. Prosperity was evident

everywhere in Christmas ads that year. Cartons of cigarettes bearing sprigs of holly and other seasonal decorations were very popular, as were electrical items of every type. Gadgets were much in vogue. My father bought my mother a hand-operated ice crusher, for creating shaved ice for cocktails, which converted perfectly good ice cubes into a small amount of cool water after twenty minutes of vigorous cranking. It was never used beyond New Year's Eve 1951, but it did grace a corner of the kitchen counter until well into the 1970s.

Tucked among the smiling ads and happy features were hints of deeper anxieties, however. *Reader's Digest* that autumn was asking 'Who Owns Your Child's Mind?' (Teachers with Communist sympathies apparently.) Polio was so rife that even *House Beautiful* ran an article on how to reduce risks for one's children. Among its tips (nearly all ineffective) were to keep all food covered, avoid sitting in cold water or wet bathing suits, get plenty of rest and, above all, be wary of 'admitting new people to the family circle'.

Harper's magazine in December struck a sombre economic note with an article by Nancy B. Mavity on an unsettling new phenomenon, the two-income family, in which husband and wife both went out to work to pay for a more ambitious lifestyle. Mavity's worry was not how women would cope with the demands of employment on top of child-rearing and housework, but rather what this would do to the man's traditional standing as breadwinner. 'I'd be ashamed to let my wife work,' one man told Mavity tartly, and it was clear from her tone that Mavity expected most readers to agree. Remarkably, until the war many women in America had been unable to work whether they wanted to or not. Up until Pearl Harbor, half of the forty-eight states had laws making it illegal to employ a married woman.

In this respect my father was commendably – I would even say enthusiastically – liberal, for there was nothing about my mother's earning capacity that didn't gladden his heart. She, too, worked for the *Des Moines Register*, as the Home Furnishings Editor, in which capacity she provided calm reassurance to two generations of homemakers who were anxious to know whether the time had come for paisley in the bedroom, whether they should have square sofa cushions or round, even whether their house itself passed muster. 'The one-story ranch house is here to stay,' she assured her readers, to presumed cries of relief in the western suburbs, in her last piece before disappearing to have me.

Because they both worked we were better off than most people of our socio-economic background (which in Des Moines in the 1950s was most people). We – that is to say, my parents, my brother Michael, my sister Mary Elizabeth (or Betty) and I – had a bigger house on a larger lot than most of my parents' colleagues. It was a white clapboard house with black shutters and a big screened porch atop a shady hill on the best side of town.

My sister and brother were considerably older than I – my sister by six years, my brother by nine – and so were effectively adults from my perspective. They were big enough to be seldom around for most of my childhood. For the first few years of my life, I shared a small bedroom with my brother. We got along fine. My brother had constant colds and allergies, and owned at least four hundred cotton handkerchiefs, which he devotedly filled with great honks and then pushed into any convenient resting place – under the mattress, between sofa cushions, behind the curtains. When I was nine he left for college and a life as a journalist in New York City, never to return permanently, and I had the room to myself after that. But I was still finding his handkerchiefs when I was in high school.

The only downside of my mother's working was that it put a little pressure on her with regard to running the home and particularly with regard to dinner, which frankly was not her strong suit anyway. My mother always ran late and was dangerously forgetful into the bargain. You soon learned to stand aside about ten to six every evening, for it was then that she would fly in the back door, throw something in the oven, and disappear into some other quarter of the house to embark on the thousand other household tasks that greeted her each evening. In consequence she nearly always forgot about dinner until a point slightly beyond way too late. As a rule you knew it was time to eat when you could hear potatoes exploding in the oven.

We didn't call it the kitchen in our house. We called it the Burns Unit.

'It's a bit burned,' my mother would say apologetically at every meal, presenting you with a piece of meat that looked like something – a much-loved pet perhaps – salvaged from a tragic house fire. 'But I think I scraped off most of the burned part,' she would add, over-looking that this included every bit of it that had once been flesh.

Happily, all this suited my father. His palate only responded to two tastes – burned and ice cream – so everything was fine by him so long as it was sufficiently dark and not too startlingly flavourful. Theirs truly was a marriage made in heaven, for no one could burn food like my mother or eat it like my dad.

As part of her job, my mother bought stacks of house-keeping magazines – *House Beautiful*, *House and Garden*, *Better Homes and Gardens*, *Good Housekeeping* – and I read these with a certain avidity, partly because they were always lying around and in our house all idle moments were spent reading something, and partly because they depicted lives so absorbingly at variance with our own. The housewives in my mother's magazines were so collected, so organized, so calmly on top of things, and their food was perfect – their *lives* were perfect. They dressed up to take their food out of the oven! There were no black circles on the ceiling above their stoves, no mutating goo climbing over the sides of their forgotten saucepans. Children didn't have to be ordered to

34

stand back every time they opened *their* oven doors. And their foods – baked Alaska, lobster Newburg, chicken cacciatore – why, these were dishes we didn't even dream of, much less encounter, in Iowa.

Like most people in Iowa in the 1950s, we were more cautious eaters in our house.* On the rare occasions when we were presented with food with which we were not comfortable or familiar – on planes or trains or when invited to a meal cooked by someone who was not herself from Iowa – we tended to tilt it up carefully with a knife and examine it from every angle as if determining whether it might need to be defused. Once on a trip to San Francisco my father was taken by friends to a Chinese restaurant and he described it to us afterwards in the sombre tones of someone recounting a near-death experience.

'And they eat it with sticks, you know,' he added knowledgeably.

'Goodness!' said my mother.

'I would rather have gas gangrene than go through that again,' my father added grimly.

In our house we didn't eat:

---

*In fact like most other people in America. The leading food writer of the age, Duncan Hines, author of the hugely successful *Adventures in Eating*, was himself a cautious eater and declared with pride that he never ate food with French names if he could possibly help it. Hines's other proud boast was that he did not venture out of America until he was seventy years old, when he made a trip to Europe. He disliked much of what he found there, especially the food.

- pasta, rice, cream cheese, sour cream, garlic, mayonnaise, onions, corned beef, pastrami, salami or foreign food of any type, except French toast;
- bread that wasn't white and at least 65 per cent air;
- spices other than salt, pepper and maple syrup;
- fish that was any shape other than rectangular and not coated in bright orange breadcrumbs, and then only on Fridays and only when my mother remembered it was Friday, which in fact was not often;
- soups not blessed by Campbell's and only a very few of those;
- anything with dubious regional names like 'pone' or 'gumbo' or foods that had at any time been an esteemed staple of slaves or peasants.

All other foods of all types – curries, enchiladas, tofu, bagels, sushi, couscous, yogurt, kale, rocket, Parma ham, any cheese that was not a vivid bright yellow and shiny enough to see your reflection in – had either not yet been invented or were still unknown to us. We really were radiantly unsophisticated. I remember being surprised to learn at quite an advanced age that a shrimp cocktail was not, as I had always imagined, a pre-dinner alcoholic drink with a shrimp in it.

All our meals consisted of leftovers. My mother had a seemingly inexhaustible supply of foods that had already been to the table, sometimes repeatedly. Apart from a few perishable dairy products, everything in the fridge was older than I was, sometimes by many years. (Her oldest food possession of all, it more or less goes without saying, was a fruit cake that was kept in a metal tin and dated from the colonial period.) I can only assume that my mother did all her cooking in the 1940s so that she could spend the rest of her life surprising herself with what she could find under cover at the back of the fridge. I never knew her to reject a food. The rule of thumb seemed to be that if you opened the lid and the stuff inside didn't make you actually recoil and take at least one staggered step backwards, it was deemed OK to eat.

Both my parents had grown up in the Great Depression and neither of them ever threw anything away if they could possibly avoid it. My mother routinely washed and dried paper plates, and smoothed out for reuse spare aluminium foil. If you left a pea on your plate, it became part of a future meal. All our sugar came in little packets spirited out of restaurants in deep coat pockets, as did our jams, jellies, crackers (oyster *and* saltine), tartare sauces, some of our ketchup and butter, all of our napkins, and a very occasional ashtray; anything that came with a restaurant table really. One of the happiest moments in my parents' life was when maple syrup started to be served in small disposable packets and they could add those to the household hoard.

Under the sink, my mother kept an enormous collection of jars, including one known as the toity jar. 'Toity' in our house was the term for a pee, and throughout my early years the toity jar was called into service whenever a need to leave the house inconveniently coincided with a sudden need by someone – and when I say 'someone', I mean of course the youngest child: me – to pee.

'Oh, you'll have to go in the toity jar then,' my mother would say with just a hint of exasperation and a worried glance at the kitchen clock. It took me a long time to realize that the toity jar was not always – or even often – the same jar twice. In so far as I thought about it at all, I suppose I guessed that the toity jar was routinely discarded and replaced with a fresh jar – we had hundreds after all.

So you may imagine my consternation, succeeded by varying degrees of dismay, when I went to the fridge one evening for a second helping of halved peaches and realized that we were all eating from a jar that had, only days before, held my urine. I recognized the jar at once because it had a Z-shaped strip of label adhering to it that uncannily recalled the mark of Zorro – a fact that I had cheerfully remarked upon as I had filled the jar with my precious bodily nectars, not that anyone had listened of course. Now here it was holding our dessert peaches. I couldn't have been more surprised if I had just been handed a packet of photos showing my mother *in flagrante* with, let's say, the guys at the gas station.

'Mom,' I said, coming to the dining-room doorway and holding up my find, 'this is the *toity* jar.'

'No, honey,' she replied smoothly without looking up. 'The toity jar's a *special* jar.'

'What's the toity jar?' asked my father with an amused air, spooning peach into his mouth.

'It's the jar I toity in,' I explained. 'And this is it.'

'Billy toities in a jar?' said my father, with very slight difficulty, as he was no longer eating the peach half he had just taken in, but resting it on his tongue pending receipt of further information concerning its recent history.

'Just occasionally,' my mother said.

My father's mystification was now nearly total, but his mouth was so full of unswallowed peach juice that he could not meaningfully speak. He asked, I believe, why I didn't just go upstairs to the bathroom like a normal person. It was a fair question in the circumstances.

'Well, sometimes we're in a hurry,' my mother went on, a touch uncomfortably. 'So I keep a jar under the sink – a special jar.'

I reappeared from the fridge, cradling more jars – as many as I could carry. 'I'm pretty sure I've used all these too,' I announced.

'That can't be right,' my mother said, but there was a kind of question mark hanging off the edge of it. Then she added, perhaps a touch self-destructively: 'Anyway, I always rinse all jars thoroughly before reuse.'

My father rose and walked to the kitchen, inclined over the waste bin and allowed the peach half to fall into it, along with about half a litre of goo. 'Perhaps a toity jar's not such a good idea,' he suggested.

So that was the end of the toity jar, though it all worked out for the best, as these things so often do. After that, all my mother had to do was mention that she had something good in a jar in the fridge and my father would get a sudden urge to take us to Bishop's, a cafeteria downtown, which was the best possible outcome, for Bishop's was the finest restaurant that ever existed.

Everything about it was divine – the food, the understated decor, the motherly waitresses in their grey uniforms who carried your tray to a table for you and gladly fetched you a new fork if you didn't like the look of the one provided. Each table had a little light on it that you could switch on if you needed service, so you never had to crane round and flag down passing waitresses. You just switched on your private beacon and after a moment a waitress would come along to see what she could help you with. Isn't that a wonderful idea?

The restrooms at Bishop's had the world's only atomic toilets – at least the only ones I have ever encountered. When you flushed, the seat automatically lifted and retreated into a seat-shaped recess in the wall, where it was bathed in a purple light that thrummed in a warm, hygienic, scientifically advanced fashion, then gently came down again impeccably sanitized, nicely warmed and practically pulsing with atomic thermoluminescence.

Goodness knows how many Iowans died from un-explained cases of buttock cancer throughout the 1950s and '60s, but it was worth every shrivelled cheek. We used to take visitors from out of town to the restrooms at Bishop's to show them the atomic toilets and they all agreed that they were the best they had ever seen.

But then most things in Des Moines in the 1950s were the best of their type. We had the smoothest, most mouth-pleasing banana cream pie at the Toddle House and I'm told the same could be said of the cheesecake at Johnny and Kay's, though my father was much too ill-at-ease with quality, and far too careful with his money, ever to take us to that outpost of fine dining on Fleur Drive. We had the most vividly delicious neon-coloured ice creams at Reed's, a parlour of cool opulence near Ashworth Swimming Pool (itself the handsomest, most elegant public swimming pool in the world, with the slimmest, tannest female lifeguards) in Greenwood Park (best tennis courts, most decorous lagoon, comeliest drives). Driving home from Ashworth Pool through Greenwood Park, under a flying canopy of green leaves, nicely basted in chlorine and knowing that you would shortly be plunging your face into three gooey scoops of Reed's ice cream is the finest feeling of well-being a person can have.

We had the tastiest baked goods at Barbara's Bake Shoppe, the meatiest, most face-smearing ribs and crispiest fried chicken at a restaurant called the Country Gentleman, the best junk food at a drive-in called George the Chilli King. (And the best farts afterwards; a George's chilli burger was gone in minutes, but the farts, it was said, went on for ever.) We had our own department stores, restaurants, clothing stores, supermarkets, drug stores, florist's, hardware stores, movie theatres, hamburger joints, you name it – every one of them the best of its kind.

Well, actually, who could say if they were the best of their kind? To know that, you'd have had to visit thousands of other towns and cities across the nation and taste all their ice cream and chocolate pie and so on because every place was different then. That was the glory of living in a world that was still largely free of global chains. Every community was special and nowhere was like everywhere else. If our commercial enterprises in Des Moines weren't the best, they were at least ours. At the very least, they all had things about them that made them interesting and different. (And they were the best.)

Dahl's, our neighbourhood supermarket, had a feature of inspired brilliance called the Kiddie Corral. This was a snug enclosure, built in the style of a cowboy corral and filled with comic books, where moms could park their kids while they shopped. Comics were produced in massive numbers in America in the 1950s – one billion of them in 1953 alone – and most of them ended up in the Kiddie Corral. It was *filled* with comic books. To enter the Kiddie Corral you climbed on to the top rail and dove in, then swam to the centre. You didn't care how long your mom took shopping because you had an infinite supply of comics to occupy you. I believe there were kids who lived in the Kiddie Corral. Sometimes when searching for the latest issue of *Rubber Man*, you would find a child buried under a foot or so of comics fast asleep or perhaps just enjoying their lovely papery smell. No institution has ever done a more thoughtful thing for children. Whoever dreamed up the Kiddie Corral is unquestionably in heaven now; he should have won a Nobel prize.

Dahl's had one other feature that was much admired. When your groceries were bagged (or 'sacked' in Iowa) and paid for, you didn't take them to your car with you, as in more mundane supermarkets, but rather you turned them over to a friendly man in a white apron who gave you a plastic card with a number on it and placed the groceries on a special sloping conveyor belt that carried them into the bowels of the earth and through a flap into a mysterious dark tunnel. You then collected your car and drove to a small brick building at the edge of the parking lot, a hundred or so feet away, where your groceries, nicely shaken and looking positively refreshed from their subterranean adventure, reappeared a minute or two later and were placed in your car by another helpful man in a white apron who took back the plastic card and wished you a happy day. It wasn't a particularly efficient system – there was often a line of cars at the little brick building if truth be told, and the juddering tunnel ride didn't really do anything except dangerously overexcite all carbonated beverages for at least two hours afterwards – but everyone loved and admired it anyway.

It was like that wherever you went in Des Moines in those days. Every commercial enterprise had something distinctive to commend it. The New Utica department store downtown had pneumatic tubes rising from each cash register. The cash from your purchase was placed in a cylinder, then inserted in the tubes and noisily fired – like a torpedo – to a central collection point, such was the urgency to get the money counted and back into the economy. A visit to the New Utica was like a trip to a future century.

Frankel's, a men's clothing store on Locust Street downtown, had a rather grand staircase leading up to a mezzanine level. A stroll around the mezzanine was a peculiarly satisfying experience, like a stroll around the deck of a ship, but more interesting because instead of looking down on empty water, you were taking in an active world of men's retailing. You could listen in on conversations and see the tops of people's heads. It had all the satisfactions of spying without any of the risks. If your dad was taking a long time being fitted for a jacket, or was busy demonstrating isometrics to the sales force, it didn't matter.

'Not a problem,' you'd call down generously from your lofty position. 'I'll do another circuit.'

Even better in terms of elevated pleasures was the Shops Building on Walnut Street. A lovely old office building some seven or eight storeys high and built in a faintly Moorish style, it housed a popular coffee shop in its lobby on the ground floor, above which rose, all the way to a distant ceiling, a central atrium, around which ran the building's staircase and galleried hallways. It was the dream of every young boy to get up that staircase to the top floor.

Attaining the staircase required cunning and a timely dash because you had to get past the coffee-shop manageress, a vicious, eagle-eyed stick of a woman named Mrs Musgrove who hated little boys (and for good reason, as we shall see). But if you selected the right moment when her attention was diverted, you could sprint to the stairs and on up to the dark eerie heights of the top floor, where you had a kind of gun-barrel view of the diners far below. If, further, you had some kind of hard candy with you – peanut M&Ms were especially favoured because of their smooth aerodynamic shape – you had a clear drop of seven or eight storeys. A peanut M&M that falls seventy feet into a bowl of tomato soup makes one *heck* of a splash, I can tell you.

You never got more than one shot because if the bomb missed the target and hit the table – as it nearly always did – it would explode spectacularly in a thousand candy-coated shards, wonderfully startling to the diners, but a call to arms to Mrs Musgrove, who would come flying up the stairs at about the speed that the M&M had gone down, giving you less than five seconds to scramble out a window and on to a fire escape and away to freedom.

Des Moines's greatest commercial institution was Younker Brothers, the principal department store downtown. Younkers was enormous. It occupied two buildings, separated at ground level by a public alley, making it the only department store I've ever known, possibly the only one in existence, where you could be run over while going from menswear to cosmetics. Younkers had an additional outpost across the street, known as the Store for Homes, which housed its furniture departments and which could be reached by means of an underground passageway beneath Eighth Street, via the white goods department. I've no idea why, but it was immensely satisfying to enter Younkers from the east side of Eighth and emerge a short while later, shopping completed, on the western side. People from out in the state used to come in specially to walk the passageway and to come out across the street and say, 'Hey. Whoa. Golly.'

Younkers was the most elegant, up to the minute, briskly efficient, satisfyingly urbane place in Iowa. It employed twelve hundred people. It had the state's first escalators – 'electric stairways' they were called in the early days – and first air conditioning. Everything about it – its silkily swift revolving doors, its gliding stairs, its whispering elevators, each with its own white-gloved operator – seemed designed to pull you in and keep you happily, contentedly consuming. Younkers was so vast and wonderfully rambling that you seldom met anyone who really knew it all. The book department inhabited a shadowy, secretive balcony area, reached by a pokey set of stairs, that made it cosy and club-like – a place known only to aficionados. It was an outstanding book department, but you can meet people who grew up in Des Moines in the 1950s who had no idea that Younkers *had* a book department.

But its *sanctum sanctorum* was the Tea Room, a place where doting mothers took their daughters for a touch of elegance while shopping. Nothing about the Tea Room remotely interested me until I learned of a ritual that my sister mentioned in passing. It appeared that young visitors were invited to reach into a wooden box containing small gifts, each beautifully wrapped in white tissue and tied with ribbon, and select one to take away as a permanent memento of the occasion. Once my sister passed on to me a present she had acquired and didn't much care for – a die-cast coach and horses. It was only two and a half inches long, but exquisite in its detailing. The doors opened. The wheels turned. A tiny driver held thin metal reins. The whole thing had obviously been hand-painted by some devoted, underpaid person from the defeated side of the Pacific Ocean. I had never seen, much less owned, such a fine thing before.

From time to time after that for years I besought them to take me with them when they went to the Tea Room, but they always responded vaguely that they didn't like the Tea Room so much any more or that they had too much shopping to do to stop for lunch. (Only years later did I discover that in fact they went every week; it was one of those secret womanly things moms and daughters did together, like having periods and being fitted for bras.) But finally there came a day when I was perhaps eight or nine that I was shopping downtown with my mom, with my sister not there, and my mother said to me, 'Shall we go to the Tea Room?'

I don't believe I have ever been so eager to accept an invitation. We ascended in an elevator to a floor I didn't even know Younkers had. The Tea Room was the most elegant place I had ever been – like a state room from Buckingham Palace magically transported to the Middle West of America. Everything about it was starched and classy and calm. There was light music of a refined nature and the tink of cutlery on china and of ice water carefully poured. I cared nothing for the food, of course. I was waiting only for the moment when I was invited to step up to the toy box and make a selection.

When that moment came, it took me for ever to decide. Every little package looked so perfect and white, so ready to be enjoyed. Eventually, I chose an item of middling size and weight, which I dared to shake lightly. Something inside rattled and sounded as if it might be die cast. I took it to my seat and carefully unwrapped it. It was a miniature doll – an Indian baby in a papoose, beautifully made but patently for a girl. I returned with it and its disturbed packaging to the slightly backward-looking fellow who was in charge of the toy box.

'I seem to have got a *doll*,' I said, with something approaching an ironic chuckle.

He looked at it carefully. 'That's surely a shame because you only git one try at the gift box.'

'Yes, but it's a *doll*,' I said. 'For a girl.'

'Then you'll just have to git you a little girl friend to give it to, won'tcha?' he answered and gave me a toothy grin and an unfortunate wink.

Sadly, those were the last words the poor man ever spoke. A moment later he was just a small muffled shriek and a smouldering spot on the carpet.

Too late he had learned an important lesson. You really should never fuck with the Thunderbolt Kid.

Chapter 2

WELCOME TO KID WORLD

DETROIT, MICH. (AP) – Great news for boys! A prominent doctor has defended a boy's right to be dirty. Dr. Harvey Flack, director of the magazine *Family Doctor*, said in the September issue: 'Boys seem to know instinctively a profound dermatological truth – that an important element of skin health is the skin's own protective layer of grease. This should not be disturbed too frequently by washing.'

– *Des Moines Register*, 28 August 1958

So THIS IS A BOOK about not very much: about being small and getting larger slowly. One of the great myths of life is that childhood passes quickly. In fact, because time moves more slowly in Kid World – five times more slowly in a classroom on a hot afternoon, eight times more slowly on any car journey of over five miles (rising to eighty-six times more slowly when driving across Nebraska or Pennsylvania lengthwise), and so slowly during the last week before birthdays, Christmases and summer vacations as to be functionally immeasurable – it goes on for decades when measured in adult terms. It is adult life that is over in a twinkling.

The slowest place of all in my corner of the youthful firmament was the large cracked leather dental chair of Dr D. K. Brewster, our spooky, cadaverous dentist, while waiting for him to assemble his instruments and get down to business. There time didn't move forward at all. It just hung.

Dr Brewster was the most unnerving dentist in America. He was, for one thing, about a hundred and eight years old and had more than a hint of Parkinsonism in his wobbly hands. Nothing about him inspired confidence. He was perennially surprised by the power of his own equipment. 'Whoa!' he'd say as he briefly enlivened some screaming device or other. 'You could do some damage with *that*, I bet!'

Worse still, he didn't believe in novocaine. He thought it dangerous and unproven. When Dr Brewster, humming mindlessly, drilled through rocky molar and found the pulpy mass of tender nerve within, it could make your toes burst out the front of your shoes.

We appeared to be his only patients. I used to wonder why my father put us through this seasonal nightmare, and then I heard Dr Brewster congratulating him one day on his courageous frugality and I understood at once, for my father was the twentieth century's cheapest man. 'There's no point in putting yourself to the danger and expense of novocaine for anything less than the whole or partial removal of a jaw,' Dr Brewster was saying.

'Absolutely,' my father agreed. Actually he said something more like 'Abmmffffmmfff,' as he had just stepped from Dr Brewster's chair and wouldn't be able to speak intelligibly for at least three days, but he nodded with feeling.

'I wish more people felt like you, Mr Bryson,' Dr Brewster added. 'That will be three dollars, please.'

Saturdays and Sundays were the longest days in Kid World. Sunday mornings alone could last for up to three months depending on season. In central Iowa for much of the 1950s there was no television at all on Sunday mornings, so generally you just sat with a bowl of soggy Cheerios watching a test pattern until WOI-TV sputtered to life some time between about 11.25 and noon – they were fairly relaxed about Sunday starts at WOI – with an episode of *Sky King*, starring the neatly kerchiefed Kirby Grant, 'America's favorite flying cowboy' (also its only flying cowboy; also the only one with reversible names). Sky was a rancher by trade, but spent most of his time cruising the Arizona skies in his beloved Cessna, *The Songbird*, spotting cattle rustlers and other earth-bound miscreants. He was assisted in these endeavours by his dimple-cheeked, pertly buttocked niece Penny, who provided many of us with our first tingly inkling that we were indeed on the road to robust heterosexuality.

Even at six years old, and even in an age as intellectually undemanding as the 1950s, you didn't have to be hugely astute to see that a flying cowboy was a fairly flimsy premise for an action series. Sky could only capture villains who lingered at the edge of grassy landing strips and to whom it didn't occur to run for it until Sky had landed, taxied to a safe halt, climbed down from the cockpit, assumed an authoritative stance and shouted: 'OK, boys, freeze!' – a process that took a minute or two, for Kirby Grant was not, it must be said, in the first flush of youth. In consequence, the series was cancelled after just a year, so only about twenty episodes were made, all practically identical anyway. These WOI tirelessly (and, one presumes, economically) repeated for the first dozen years of my life and probably a good deal beyond. Almost the only thing that could be said in their favour was that they were more diverting than a test pattern.

The illimitable nature of weekends was both a good and a necessary thing because you always had such a lot to do in those days. A whole morning could be spent just getting the laces on your sneakers right since all sneakers in the 1950s had over seven dozen lace holes and the laces were fourteen feet long. Each morning you would jump out of bed to find that the laces had somehow become four feet longer on one side of the shoe than the other. Quite how sneakers did this just by being left on the floor overnight was a question that could not be answered – it was one of those things, like nuns and bad weather, that life threw at you periodically – but it took endless reserves of patience and scientific judgement to get them right, for no matter how painstakingly you shunted the laces around the holes, they always came out at unequal lengths. In fact, the more carefully you shunted, the more unequal they generally became. When by some miracle you finally got them exactly right, the second lace would always snap, leaving you to sigh and start again.

The makers of sneakers also thoughtfully pocked the soles with numberless crevices, craters, chevrons, mazes, crop circles and other rubbery hieroglyphs, so that when you stepped in a moist pile of dog shit, as you most assuredly did within three bounds of leaving the house, they provided additional absorbing hours of pastime while you cleaned them out with a stick, gagging quietly, but oddly content.

Hours more of weekend time needed to be devoted to picking burrs off socks, taking corks out of bottle caps, peeling frozen wrappers off Popsicles, prising apart Oreo cookies without breaking either chocolate disc half or disturbing the integrity of the filling, and carefully picking labels off jars and bottles for absolutely no reason.

In such a world, injuries and other physical setbacks were actually welcomed. If you got a splinter you could pass an afternoon, and attract a small devoted audience, seeing how far you could insert a needle under your skin – how close you could get to actual surgery. If you got sunburned you looked forward to the moment when you could peel off a sheet of translucent epidermis that was essentially the size of your body. Scabs in Kid World were cultivated the way older people cultivate orchids. I had knee scabs that I kept for up to four years, that were an inch and three quarters thick and into which you could press drawing pins without rousing my attention. Nosebleeds were much admired, needless to say, and anyone with a nosebleed was treated like a celebrity for as long as it ran.

Because days were so long and so little occurred, you were prepared to invest extended periods in just sitting and watching things on the off chance that something diverting might take place. For years, whenever my father announced that he was off to the lumberyard I dropped everything to accompany him in order to sit quietly on a stool in the wood-cutting room in the hope that Moe, the man who trimmed wood to order on a big buzzsaw, would send one of his few remaining digits flying. He had already lost most of six or seven fingers, so the chances of a lively accident always seemed good.

Buses in Des Moines in those days were electrically powered, and drew their energy from a complicated cat's cradle of overhead wires, to which each was attached by means of a metal arm. Especially in damp weather, the wires would spark like fireworks at a Mexican fiesta as the arm rubbed along them, vivdly underscoring the murderous potency of electricity. From time to time, the bus-arm would come free of the wires and the driver would have to get out with a long pole and push it back into place – an event that I always watched with the keenest interest because my sister assured me that there was every chance he would be electrocuted.

Other long periods of the day were devoted to just seeing what would happen – what would happen if you pinched a matchhead while it was still hot or made a vile drink and took a sip of it or focused a white-hot beam of sunlight with a magnifying glass on your Uncle Dick's bald spot while he was napping. (What happened was that you burned an amazingly swift, deep hole that would leave Dick and a team of specialists at Iowa Lutheran Hospital puzzled for weeks.)

Thanks to such investigations and the abundance of

time that made them possible, I knew more things in the first ten years of my life than I believe I have known at any time since. I knew everything there was to know about our house for a start. I knew what was written on the undersides of tables and what the view was like from the tops of bookcases and wardrobes. I knew what was to be found at the back of every closet, which beds had the most dust-balls beneath them, which ceilings the most interesting stains, and where exactly the patterns in wallpaper repeated. I knew how to cross every room in the house without touching the floor, where my father kept his spare change and how much you could safely take without his noticing (one-seventh of the quarters, one-fifth of the nickels and dimes, as many of the pennies as you could carry). I knew how to relax in an armchair in more than one hundred positions and on the floor in approximately seventy-five more. I knew what the world looked like when viewed through a Jell-O lens. I knew how things tasted – damp washcloths, pencil ferrules, coins and buttons, almost anything made of plastic that was smaller than, say, a clock radio, mucus of every variety of course – in a way that I have more or less forgotten now. I knew and could take you at once to any illustration of naked women anywhere in our house, from a Rubens painting of fleshy chubbos in *Masterpieces of World Painting* to a cartoon by Peter Arno in the latest issue of the *New Yorker* to my father's small private library of girlie magazines in a secret place, known only to him, me and one hundred and eleven of my closest friends, in his bedroom.

I knew how to get between any two properties in the neighbourhood, however tall the fence or impenetrable the hedge that separated them. I knew the feel of linoleum on bare skin and what everything smelled like at floor level. I knew pain the way you know it when it is fresh and interesting – the pain, for example, of a toasted marsh-mallow in your mouth when its interior is roughly the temperature of magma. I knew exactly how clouds drifted on a July afternoon, what rain tasted like, how ladybirds preened and caterpillars rippled, what it felt like to sit inside a bush. I knew how to appreciate a really good fart, whether mine or someone else's.

The someone else was nearly always Buddy Doberman, who lived across the alley, a secretive lane that ran in a neighbourly fashion behind our houses. Buddy was my best friend for the first portion of my life. We were extremely close. He was the only human being whose anus I have ever looked at closely, or indeed at all, just to see what one looks like (reddish, tight and very slightly puckered, as I recall with a rather worrying clarity) and he was good tempered and had wonderful toys to play with, as his parents were both generous and well to do.

He was sweetly stupid, too, which was a bonus. When he and I were four his grandfather gave us a pair of wooden pirate swords that he had made in his workshop and we went with them more or less straight to Mrs Van

Pelt's prized flower border, which ran for about thirty yards along the alley. In a whirl of frenzied motion that anticipated by several years the lively destructive actions of a strimmer, we decapitated and eviscerated every one of her beloved zinnias in a matter of seconds. Then, realizing the enormity of what we had just done – Mrs Van Pelt showed these flowers at the state fair; she talked to them; they were her children – I told Buddy that this was not a good time for me to be in trouble on account of my father had a fatal disease that no one knew about, so would he mind taking all the blame? And he did. So while he was sent to his room at three o'clock in the afternoon and spent the rest of the day as a weepy face at a high window, I was on our back porch with my feet up on the rail, gorging on fresh watermelon and listening to selected cool discs on my sister's portable phonograph. From this I learned an important lesson: lying is always an option worth trying. I spent the next six years blaming Buddy for everything bad that happened in my life. I believe he eventually even took the rap for burning the hole in my Uncle Dick's head even though he had never met my Uncle Dick.

Then, as now, Des Moines was a safe, wholesome city of two hundred thousand people. The streets were long, straight, leafy and clean and had solid Middle American names: Woodland, University, Pleasant, Grand. (There was a local joke, much retold, about a woman who was goosed on Grand and thought it was Pleasant.) It was a

61

nice city – a comfortable city. Most businesses were close to the road and generally had lawns out front instead of parking lots. Public buildings – post offices, schools, hospitals – were always stately and imposing. Gas stations often looked like little cottages. Diners (or roadhouses) brought to mind the type of cabins you might find on a fishing trip. Nothing was designed to be particularly helpful or beneficial to cars. It was a greener, quieter, less intrusive world.

Grand Avenue was the main artery through the city, linking downtown, where everyone worked and did all serious shopping, with the residential areas beyond. The best houses in the city lay to the south of Grand on the west side of town, in a hilly, gorgeously wooded district that ran down to Waterworks Park and the Raccoon River. You could walk for hours along the wandering roads in there and never see anything but perfect lawns, old trees, freshly washed cars and lovely, happy homes. It was miles and miles of the American dream. This was my district. It was known as South of Grand.

The most striking difference between then and now was how many kids there were then. America had thirty-two million children aged twelve or under in the mid-1950s, and four million new babies were plopping on to the changing mats every year. So there were kids everywhere, all the time, in densities now unimaginable, but especially whenever anything interesting or unusual happened. Early every summer, at the start of the mosquito season, a city employee in an open jeep would come to the neighbourhood and drive madly all over the place – over lawns, through woods, bumping along culverts, jouncing into and out of vacant lots – with a fogging machine that pumped out dense, colourful clouds of insecticide through which at least eleven thousand children scampered joyously for most of the day. It was awful stuff – it tasted foul, it made your lungs chalky, it left you with a powdery saffron pallor that no amount of scrubbing could eradicate. For years afterwards whenever I coughed into a white handkerchief I brought up a little ring of coloured powder.

But nobody ever thought to stop us or suggest that it was perhaps unwise to be scampering through choking clouds of insecticide. Possibly it was thought that a generous dusting of DDT would do us good. It was that kind of age. Or maybe we were just considered expendable because there were so many of us.*

The other difference from those days was that kids were always outdoors – I knew kids who were pushed out the back door at eight in the morning and not allowed back in until five unless they were on fire or actively bleeding – and they were always looking for something to do. If you stood on any corner with a bike – any corner

*Altogether the mothers of post-war America gave birth to 76.4 million kids between 1946 and 1964, when their poor old overworked wombs all gave out more or less at once, evidently.

anywhere – over a hundred children, many of whom you had never seen before, would appear and ask you where you were going.

'Might go down to the Trestle,' you would say thoughtfully. The Trestle was a railway bridge over the Raccoon River, from which you could jump in for a swim if you didn't mind paddling around among dead fish, old tyres, oil drums, algal slime, heavy metal effluents and uncategorized goo. It was one of ten recognized landmarks in our district. The others were the Woods, the Park, the Little League Park (or 'the Ballpark'), the Pond, the River, the Railroad Tracks (usually just 'the Tracks'), the Vacant Lot, Greenwood (our school), and the New House. The New House was any house under construction and so changed regularly.

'Can we come?' they'd say.

'Yeah, all right,' you would answer if they were your size or 'If you think you can keep up' if they were smaller. And when you got to the Trestle or the Vacant Lot or the Pond there would already be six hundred kids there. There were always six hundred kids everywhere except where two or more neighbourhoods met – at the Park, for instance – where the numbers would grow into the thousands. I once took part in an ice hockey game at the lagoon in Greenwood Park that involved four thousand kids, all slashing away violently with sticks, and went on for at least three quarters of an hour before anyone realized that we didn't have a puck.

Life in Kid World, wherever you went, was un-supervised, unregulated and robustly – at times insanely – physical and yet it was a remarkably peaceable place. Kids' fights never went too far, which is extraordinary when you consider how ill-controlled children's tempers are. Once when I was about six, I saw a kid throw a rock at another kid, from quite a distance, and it bounced off the target's head (quite beautifully, I have to say) and made him bleed. This was talked about for years. Kids in the next county knew about it. The kid who did it was sent for about ten thousand hours of therapy.

We didn't otherwise engage in violence, other than accidentally, though we did sometimes (actually quite routinely) give a boy named Milton Milton knuckle rubs for having such a stupid name and also for spending his life pretending to be motorized. I never knew whether he was supposed to be a train or a robot or what, but he always moved his arms like pistons when he walked and made puffing noises, and so naturally we gave him knuckle rubs. We had to. He was born to be rubbed.

With respect to accidental bloodshed, it is my modest boast that I became the neighbourhood's most memor-able contributor one tranquil September afternoon in my tenth year while playing football in Leo Collingwood's back yard. As always, the game involved about a hundred and fifty kids, so normally when you were tackled you fell into a soft, marshmallowy mass of bodies. If you were really lucky you landed on Mary O'Leary and got to rest on her for a moment while waiting for the others to get off. She smelled of vanilla – vanilla and fresh grass – and was soft and clean and painfully pretty. It was a lovely moment. But on this occasion I fell outside the pack and hit my head on a stone retaining wall. I remember feeling a sharp pain at the top of my head towards the back.

When I stood up, I saw that everyone was staring at me with a single rapt expression and inclined to give me some space. Lonny Brankovich looked over and instantly melted in a faint. In a candid tone his brother said: 'You're gonna die.' Naturally I couldn't see what absorbed them, but I gather from later descriptions that it looked as if I had a lawn sprinkler plugged into the top of my head, spraying blood in all directions in a rather festive manner. I reached up and found a mass of wetness. To the touch, it felt more like the kind of outflow you get when a truck crashes into a fire hydrant or oil is struck in Oklahoma. This felt like a job for Red Adair.

'I think I'd better go get this seen to,' I said soberly and with a fifty-foot stride left the yard. I bounded home in three steps and walked into the kitchen, fountaining lavishly, where I found my father standing by the window with a cup of coffee dreamily admiring Mrs Bukowski, the young housewife from next door. Mrs Bukowski had the first bikini in Iowa and wore it while hanging out her wash. My father looked at my spouting head, allowed himself a moment's mindless adjustment, then leaped instantly and adroitly into panic and disorder, moving in as many as six directions simultaneously, and calling in a strained voice to my mother to come at once and bring lots and lots of towels – 'old ones!' – because Billy was bleeding to death in the kitchen.

Everything after that went by in a blur. I remember my father seating me with my head pressed to the kitchen table as he endeavoured to staunch the flow of blood and at the same time get through on the phone to Dr Alzheimer, the family physician, for guidance. Meanwhile, my mother, ever imperturbable, searched methodically for old rags and pieces of cloth that could be safely sacrificed (or were red already) and dealt with the parade of children who were turning up at the back door with bone chips and bits of grey tissue that they had carefully lifted from the rock and thought might be part of my brain.

I couldn't see much, of course, with my head pressed to the table, but I did catch reflected glimpses in the toaster and my father seemed to be into my cranial cavity up to his elbows. At the same time he was speaking to Dr Alzheimer in words that failed to soothe. 'Jesus Christ, doc,' he was saying. 'You wouldn't *believe* the amount of blood. We're *swimming* in it.'

On the other end I could hear Dr Alzheimer's dementedly laid-back voice. 'Well, I *could* come over, I suppose,' he was saying. 'It's just that I'm watching an *awfully* good golf tournament. Ben Hogan is having a most marvellous round. Isn't it wonderful to see him doing well at his time of life? Now then, have you managed to stop the bleeding?'

'Well, I'm sure trying.'

'Good, good. That's excellent – that's excellent. Because he's probably lost quite a lot of blood already. Tell me, is the little fellow still breathing?'

'I think so,' my father replied.

I nodded swiftly and helpfully.

'Yes, he's still breathing, doc.'

'That's good, that's very good. OK, I tell you what. Give him two aspirin and nudge him once in a while to make sure he doesn't pass out – on no account let him lose consciousness, do you hear, because you might lose the poor little fellow – and I'll be over after the tournament. Oh, look at that – he's gone straight off the green into the rough.' There was the sound of Dr Alzheimer's phone settling back into the cradle and the buzz of disconnection.

Happily, I didn't die and four hours later was to be found sitting up in bed, head extravagantly turbaned, well rested after a nap that came during one of those passing three-hour moments when my parents forgot to check on my wakefulness, eating tubs of chocolate ice cream, and regally receiving visitors from the neighbourhood, giving particular priority to those who came bearing gifts. Dr Alzheimer arrived later than promised, smelling lightly of bourbon. He spent most of the visit sitting on the edge of my bed and asking me if I was old enough to remember Bobby Jones. He never did look at my head. Dr

Alzheimer's fees, I believe, were very reasonable, too.

Apart from medical practitioners, Iowa offered little in the way of natural dangers, though one year when I was about six we had an infestation of a type of giant insect called cicada killers. Cicada killers are not to be confused with cicadas, which are themselves horrible things – like small flying cigars, but with staring red eyes and grotesque pincers, if I recall correctly. Well, cicada killers were much worse. They only emerged from the ground every seventeen years, so nobody, even adults, knew much about them. There was great debate over whether the 'killer' in 'cicada killer' signified that they were killers of cicadas or that they were cicadas that killed. The consensus pointed to the latter.

Cicada killers were about the size of hummingbirds and had vicious stingers fore and aft, and they were awful. They lived in burrows and would come flying up un-expectedly from below, with a horrible whirring sound, rather like a chainsaw starting up, if you disturbed their nests. The greatest fear was that they would shoot up the leg of your shorts and become entangled in your under-pants and start lashing out blindly. Castration, possibly by the side of the road, was the normal emergency procedure for cicada killer stings to the scrotal region – and they seldom stung anywhere else. You never really saw one because as soon as one whirred out of its burrow you pranced away like hell, pressing your shorts primly but prudently to your legs.

The worst chronic threat we had was poison sumac, though I never knew anyone, adult or child, who actually knew what it was or precisely how it would kill you. It was really just a kind of shrubby rumour. Even so, in any wooded situation you could always hold up a hand and announce gravely: 'We'd better not go any further. I think there might be sumac up ahead.'

'*Poison* sumac?' one of your younger companions would reply, eyes wide open.

'All sumac's poisonous, Jimmy,' someone else would say, putting a hand on his shoulder.

'Is it really bad?' Jimmy would ask.

'Put it this way,' you would answer sagely. 'My brother's friend Mickey Cox knew a guy who fell into a sumac patch once. Got it all over him, you know, and the doctors had to like amputate his whole body. He's just a head on a plate now. They carry him around in a hatbox.'

'Wow,' everyone would say except Arthur Bergen, who was annoyingly brainy and knew all the things in the world that couldn't possibly be so, which always exactly coincided with all the things you had heard about that were amazing.

'A head couldn't survive on its own in a box,' he would say.

'Well, they took it out sometimes. To give it air and let it watch TV and so on.'

'No, I mean it couldn't survive on its own, without a body.'

'Well, this one did.'

'Not possible. How are you going to keep a head oxygenated without a heart?'

'How should I know? What am I – Dr Kildare? I just know it's true.'

'It can't be, Bryson. You've misheard – or you're making it up.'

'Well, I'm not.'

'Must be.'

'Well, Arthur, I swear to God it's true.'

This would cause an immediate stunned silence.

'You'll go to hell for saying that if it's not true, you know,' Jimmy would point out, but quite unnecessarily for you knew this already. All kids knew this automatically, from birth.

Swearing to God was the ultimate act. If you swore to God and it turned out you were wrong, even by accident, even just a little, you still had to go to hell. That was just the rule and God didn't bend that rule for anybody. So the moment you said it, in any context, you began to feel uneasy in case some part of it turned out to be slightly incorrect.

'Well, that's what my brother said,' you would say, trying to modify your eternal liability.

'You can't change it now,' Bergen – who, not incidentally, would grow up to be a personal injury lawyer – would point out. 'You've already said it.'

You were all too well aware of this too. In the

71

circumstances there was really only one thing to do: give Milton Milton a knuckle rub.

Only slightly less threatening than poison sumac were pulpy red berries that grew in clumps on bushes in almost everybody's back yard. These were also slightly vague in that neither bush nor berry seemed to have a name – they were just 'those red berries' or 'that bush with the red berries' – but they were universally agreed to be toxic. If you touched or held a berry even briefly and then later ate a cookie or sandwich and realized that you hadn't washed your hands, you spent an hour seriously wondering if you might drop dead at any moment.

Moms worried about the berries, too, and were forever shouting from the kitchen window not to eat them, which was actually unnecessary because children of the 1950s didn't eat anything that grew wild – in fact, didn't eat anything at all unless it was coated in sugar, endorsed by a celebrity athlete or TV star and came with a free prize. They might as well have told us not to eat any dead cats we found. We weren't about to.

Interestingly, the berries weren't poisonous at all. I can say this with some confidence because we made Lanny Kowalski's little brother, Lumpy,* eat about four pounds of them to see if they would kill him and they didn't. It was a controlled experiment, I hasten to add. We fed them to him one at a time and waited a decent interval

*So called because his pants always had a saggy lump of poo in them. I expect they still do.

to see if his eyes rolled up into his head or anything before passing him another. But apart from throwing up the middle two pounds, he showed no ill effects.

The only real danger in life was the Butter boys. The Butters were a family of large, inbred, indeterminately numerous individuals who lived seasonally in a collection of shanty homes in an area of perpetual wooded gloom known as the Bottoms along the swampy margins of the Raccoon River. Nearly every spring the Bottoms would flood and the Butters would all go back to Arkansas or Alabama or wherever it was they came from.

In between times they would menace us. Their speciality was to torment any children smaller than them, which was all children. The Butters were big to begin with but because they were held back year after year, they were much, much larger than any other child in their class. By sixth grade some of them were too big to pass through doors. They were ugly, too, and real dumb. They ate squirrels.

Generally the best option was to have some small child that you could offer as a sacrifice. Lumpy Kowalski was ideal for this as he was indifferent to pain and fear, and would never tell on you because he couldn't, or possibly just didn't, speak. (It was never clear which.) Also, the Butters were certain to be grossed out by his dirty pants, so they would merely paw him for a bit and then withdraw with pained, confused faces.

The worst outcome was to be caught on your own by

one or more of the Butter boys. Once when I was about ten I was nabbed by Buddy Butter, who was in my grade but at least seven years older. He dragged me under a big pine tree and pinned me to the ground on my back and told me he was going to keep me there all night long.

I waited for what seemed a decent interval and then said: 'Why are you doing this to me?'

'Because I can,' he answered, but pronounced it 'kin'. Then he made a kind of glutinous, appreciative, snot-clearing noise, which was what passed in the Butter universe for laughter.

'But you'll have to stay here all night, too,' I pointed out. 'It'll be just as boring for you.'

'Don't care,' he replied, sharp as anything, and was quiet a long time before adding: 'Besides I can do this.' And he treated me to the hanging spit trick – the one where the person on top slowly suspends a gob of spit and lets it hang there by a thread, trembling gently, and either sucks it back in if the victim surrenders or lets it fall, sometimes inadvertently. It wasn't even like spit – at least not like human spit. It was more like the sort of thing a giant insect would regurgitate on to its forelimbs and rub on to its antennae. It was a mossy green with little streaks of red blood in it and, unless my memory is playing tricks, two very small grey feathers protruding at the sides. It was so big and shiny that I could see my reflection in it, distorted, as in an M. C. Escher drawing. I knew that if any part of it

74

touched my face, it would sizzle hotly and leave a dis-figuring scar.

In fact, he sucked the gob back in and got off me. 'Well, you let that be a lesson to you, you little skunk pussy poontang sissy,' he said.

Two days later the soaking spring rains came and put all the Butters on their tarpaper roofs, where they were res-cued one by one by men in small boats. A thousand children stood on the banks above and cheered.

What they didn't realize was that the storm clouds that carried all that refreshing rain had been guided across the skies by the powerful X-ray vision of the modest super-hero of the prairies, the small but perfectly proportioned Thunderbolt Kid.

Chapter 3

BIRTH OF A SUPERHERO

EAST HAMPTON, CONN. (AP) – A search of Lake Pocotopaug for a reported drowning victim was called off here Tuesday when it was realized that one of the volunteers helping the search, Robert Hausman, 23, of East Hampton, was the person being sought.

– *Des Moines Register*, 20 September 1957

AT EVERY MEAL she ever prepared throughout my upbringing (and no doubt far beyond), my mother placed a large dollop of cottage cheese on each plate. It appeared to be important to her to serve something co-agulated and slightly runny at every meal. It would be understating things to say I disliked cottage cheese. To me cottage cheese looks like something you bring up, not take in. Indeed, that was the crux of my problem with it.

I had a distant uncle named Dee (who, now that I think of it, may not have actually been an uncle at all, but just a strange man who showed up at all large family gatherings) who had lost his voicebox and had a perma-nent hole in his throat as a result of some youthful injury or surgical trauma or something. Actually, I don't know why he had a hole in his throat. It was just a fact of life. A lot of rural people in Iowa in the Fifties had arresting physical features – wooden legs, stumpy arms, outstand-ingly dented heads, hands without fingers, mouths without tongues, sockets without eyes, scars that ran on for feet, sometimes going in one sleeve and out the other. Goodness knows what people got up to back then, but they suffered some mishaps, that's for sure.

Anyway, Uncle Dee had a throat hole, which he kept lightly covered with a square of cotton gauze. The gauze often came unstuck, particularly when Dee was in an impassioned mood, which was usually, and either hung open or fell off altogether. In either case, you could see the hole, which was jet black and transfixing and about the size of a quarter. Dee talked through the hole in his neck – actually, belched a form of speech through it. Everyone agreed that he was very good at it – in terms of volume and steadiness of output, he was a wonder; many were reminded of an outboard motor running at full throttle – though in fact no one had the faintest idea what he was talking about, which was unfortunate as Dee was ferociously loquacious. He would burp away with feeling while those people standing beside him (who were, it must be said, nearly always newcomers to the family circle) watched his throat hole gamely but uncertainly. From time to time, they would say, 'Is that so?' and 'Well, I'll be,' and give a series of earnest, thoughtful nods, before saying, 'Well, I think I'll just go and get a little more lemonade,' and drift off, leaving Dee belching furiously at their backs.

All this was fine – or at least fine enough – so long as Uncle Dee wasn't eating. When Dee was eating you really didn't want to be in the same county, for Uncle Dee talked with his throat full. Whatever he ate turned into a light spray from his throat hole. It was like dining with a miniature flocking machine, or perhaps a very small snowblower. I've seen placid, kindly grown-ups, people of good Christian disposition – loving sisters, sons and fathers, and on one memorable occasion two Lutheran ministers from neighbouring congregations – engage in silent but prolonged and ferocious struggles for control of a chair that would spare them having to sit beside or, worse, across from Dee at lunch.

The feature of Dee's condition that particularly caught my attention was that whatever he put in his mouth – chocolate cream pie, chicken-fried steak, baked beans, spinach, rutabaga, Jell-O – by the time it reached the hole in his neck it had become cottage cheese. I don't know how but it did.

Which is precisely and obviously why I disdained the stuff. My mother could never grasp this. But then she was dazzlingly, good-naturedly forgetful about most things. We used to amuse ourselves by challenging her to supply our dates of birth or, if that proved too taxing, the seasons. She couldn't reliably tell you our middle names. At the supermarket she often reached the checkout and discovered that she had at some indeterminate point acquired someone else's shopping cart, and was now in possession of items – whole pineapples, suppositories, bags of food for a very large dog – that she didn't want or mean to have. She was seldom entirely clear on what clothes belonged to whom. She hadn't the faintest idea what our eating preferences were.

'Mom,' I would say each night, laying a piece of bread

over the offending mound on my plate, rather as one covered a roadside accident victim with a blanket, 'you know I really do hate cottage cheese.'

'Do you, dear?' she would say with a look of sympathetic perplexity. 'Why?'

'It looks like the stuff that comes out of Uncle Dee's throat.'

Everyone present, including my father, would nod solemnly at this.

'Well, just eat a little bit, and leave what you don't like.'

'I don't like any of it, Mom. It's not like there's a part of it I like and a part that I don't. Mom, we have this conversation every night.'

'I bet you've never even tasted it.'

'I've never tasted pigeon droppings. I've never tasted ear wax. Some things you don't need to taste. We have this conversation every night, too.'

More solemn nods.

'Well, I had no idea you didn't like cottage cheese,' my mom would say in something like amazement and the next night there would be cottage cheese again.

Just occasionally her forgetfulness strayed into rather more dismaying territory, especially when she was pressed for time. I recall one particularly rushed and disorganized morning when I was still quite small – small enough, at any rate, to be mostly trusting and completely stupid – when she gave me my sister's old Capri pants to wear to school. They were a brilliant lime green, very tight, and had little slits at the bottom. They only came about three quarters of the way down my calves. I stared at myself in the back hall mirror in a kind of confused disbelief. I looked like Barbara Stanwyck in *Double Indemnity*.

'This can't be right, Mom,' I said. 'These are Betty's old Capri pants, aren't they?'

'No, honey,' my mom replied soothingly. 'They're *pirate* pants. They're very fashionable. I believe Kookie Byrnes wears them on *77 Sunset Strip*.'

Kookie Byrnes, a munificently coiffed star on this popular weekly television show, was a hero to me, and indeed to most people who liked interestingly arranged hair, and he *was* capable of endearingly strange things, that's for sure. That's why they called him Kookie. Even so, this didn't feel right.

'I don't think he can, Mom. Because these are girls' pants.'

'He does, honey.'

'Do you swear to God?'

'Oom,' she said distractedly. 'You watch this week. I'm sure he does.'

'But do you swear to God?'

'Oom,' she said again.

So I wore them to school and the laughter could be heard for miles. It went on for most of the day. The principal, Mrs Unnaturally Enormous Bosom, who in normal circumstances was the sort of person who wouldn't get off her ass if her chair was on fire, made a special visit to have a look at me and laughed so hard she popped a button on her blouse.

Kookie Byrnes, of course, never wore anything remotely like Capri pants. I asked my sister about this after school. 'Are you *kidding*?' she said. 'Kookie Byrnes is *not* homosexual.'

It was impossible to hold my mother's forgetfulness against her for long because it was so obviously and help-lessly pathological, a quirk of her nature. We might as well have become exasperated with her for having a fondness for polka dots and two-tone shoes. It's what she was. Besides, she made up for it in a thousand ways – by being soft and kind, patient and generous, instantly and sincerely apologetic for every wrong, keen to make amends. Everybody in the world adored my mother. She was entirely without suspicion or malice. She never raised her voice or said no to any request, never said a word against any other human being. She liked everybody. She lived to make sandwiches. She wanted everyone to be happy. And she took me almost every week to dinner and the movies. It was the thing she and I did together.

Because of his work my father was gone most weekends, so every Friday, practically without fail, my mother would say to me, 'What do you say we go to Bishop's for dinner tonight and then take in a movie?' as if it were a rare treat, when in fact it was what we did regularly.

So at the conclusion of school on Fridays I would hasten home, drop my books on the kitchen table, grab a handful of cookies and proceed downtown. Sometimes I caught a bus, but more often I saved the money and walked. It was only a couple of miles and the route was all diverting and agreeable if I went along Grand Avenue (where the buses didn't go; they were relegated to Ingersoll – the servants' entrance of the street world). I liked Grand Avenue very much. In those days it was adorned from downtown to the western suburbs with towering, interlaced elms, the handsomest streetside tree ever and a generous provider of drifts of golden leaves to shuffle through in autumn. But more than this, Grand felt the way a street should feel. Its office buildings and apartments were built close to the road, which gave the street a kind of neighbourliness, and it still had most of its old homes – mansions of exuberant splendour, nearly all with turrets and towers and porches like ships' decks – though these had now mostly found other uses as offices, funeral homes and the like. Interspersed at judicious intervals were a few grander institutional buildings: granite churches, a Catholic girls' high school, the stately Commodore Hotel (with an awninged walkway leading to the street – a welcome touch of Manhattan), a spooky orphanage where no children ever played or stood at a window, the official residence of the governor, a modest mansion with a white flagpole and the state flag. All seemed somehow exactly in proportion, precisely positioned, thoughtfully dressed and groomed. It was the perfect street.

Where it ceased being residential and entered the downtown, by the industrial-scale hulk of the Meredith Publishing Building (home of *Better Homes and Gardens* magazine), Grand made an abrupt dog leg to the left, as if it suddenly remembered an important appointment. Originally from this point it was intended to proceed through the downtown as a kind of Midwestern Champs Elysées, running up to the steps of the state capitol building. The idea was that as you progressed along Grand you would behold before you, perfectly centred, the golden-domed glory of the capitol building (and it *is* quite a structure, one of the best in the country).

But when the road was being laid out some time in the second half of the 1800s there was a heavy rain in the night and the surveyors' sticks moved apparently – at least that was what we were always told – and the road deviated from the correct line, leaving the capitol oddly off centre, so that it looks today as if it has been caught in the act of trying to escape. It is a peculiarity that some people treasure and others would rather not talk about. I for one never tired of striding into the downtown from the west and being confronted with a view so gloriously not right, so cherishably out of kilter, and pondering the fact that whole teams of men could build an important road evidently without once looking up to see where they were going.

For its first couple of blocks, Des Moines's downtown had a slight but agreeably seedy air. Here there were dark bars, small hotels of doubtful repute, dingy offices and shops that sold odd things like rubber stamps and trusses. I liked this area very much. There was always a chance of hearing a bitter argument through an upstairs window and the hope that this would lead to gunfire and someone falling out the window on to an awning, as in the better Hollywood movies, or at least staggering out a door, hand on bloodied chest, and collapsing in the street.

Then fairly quickly downtown became more respectable and literally upstanding, more like a real downtown. This throbbing heart of the metropolis was of a fairly modest scale – only three or four blocks wide and four or five long – but it had a density of tallish brick buildings and it was full of people and life. The air was slightly dirty and blue. People walked with quicker steps and longer strides. It felt like a proper city.

Upon arriving downtown I had an unvarying routine. I would call first at Pinky's, a joke and novelty store in the Banker's Trust Building, which contained a large stock of dusty gag items – plastic ice cubes with a fly inside, chattering teeth, rubber turds for every occasion – that no one ever bought. Pinky's existed purely to give sailors, migrant workers and small boys a place to go when they were at a loose end downtown. I have no idea how it managed to stay in business. I can only assume that somehow in the 1950s you didn't have to sell much to remain solvent.*

When I had looked at everything there, I would do a circuit or two of the mezzanine at Frankel's, then check out any new Hardy Boys arrivals in the book department at Younkers. Generally I would call in at the long soda fountain at Woolworth's for one of their celebrated Green Rivers, a refreshing concoction of syrupy green fizz that was the schoolboy aperitif of the 1950s, and finally head over to the R&T (for *Register* and *Tribune*), at Eighth and Locust. There I would always take a minute to look through the large plate-glass windows that ran along the building at street level and gave views of the press room – a potentially excellent place to see a mangling, I always supposed – then proceed through the snappy revolving doors into the *Register*'s lobby, where I would devote a few respectful minutes to a large, slowly revolving globe that was housed behind glass (always interestingly warm to the touch) in a side room.

The *Register* was proud of this globe. It was, as I recall, one of the largest globes in the world: big globes aren't easy to make apparently. This one was at least twice my size and beautifully manufactured and painted. It was tilted on its axis at a scientifically precise angle and spun

---

*I have since learned from my more worldly informant Stephen Katz that Pinky's earned its keep by selling dirty magazines under the counter. I had no idea.

at the same speed as Earth itself, completing one revolution in every twenty-four hours. It was, in short, a thing of wonder and grandeur – the finest technological marvel in Des Moines aside from the radioactive toilet seats at Bishop's cafeteria, which obviously were in a league of their own. Because it was so large and stately and real, it felt very much as if you were looking at the actual Earth, and I would walk around it imagining myself as God. Even now when I think of the nations of the Earth, I see them as they were on that big ball – as Tanganyika, Rhodesia, East and West Germany, the Friendly Islands. The globe may have had other fans besides me, but I never saw any passer-by give it so much as a glance.

At 5.30 precisely, I would proceed in an elevator up to the newsroom on the fourth floor – a place so quint-essentially like a newsroom that it even had a swing gate through which you entered with a jaunty air, like Rosalind Russell in *His Girl Friday* – and passed through the Sports Department with a familiar 'Hey' to all the fellows there (they were my father's colleagues after all), past the chattering wire machines, and presented myself at my mother's desk in the Women's Department just beyond. I can see her perfectly now, sitting at a grey metal desk, hair slightly askew, hammering away on her typewriter, a venerable Smith Corona upright. I'd give anything – really almost anything at all – to pass just once more through that gate and see the guys in the Sports Department and beyond them my dear old mom at her desk typing away.

My arrival would always please and surprise her equally. 'Why, Billy, hello! My goodness, is it Friday?' she would say as if we hadn't met for weeks.

'Yes, Mom.'

'Well, what do you say we go to Bishop's and a movie?'

'That would be great.'

So we would dine quietly and contentedly at Bishop's and afterwards stroll to a movie at one of the three great and ancient downtown movie palaces – the Paramount, Des Moines and RKO-Orpheum – each a vast, spookily lit crypt done up in an elaborate style that recalled the heyday of ancient Egypt. The Paramount and Des Moines both held sixteen hundred people, the Orpheum slightly fewer, though by the late 1950s there were seldom more than thirty or forty at a showing. There has never been, will never again be, a better place to pass a Friday evening, sitting with a tub of buttered popcorn in a cubic acre of darkness facing a screen so enormous you could read the titles of books on bookcases, the dates on calendars, the licence plates of passing cars. It really was a kind of magic.

Movies of the Fifties were of unparalleled excellence. *The Brain That Wouldn't Die*, *The Blob*, *The Man from Planet X*, *Earth Versus the Flying Saucers*, *Zombies of the Stratosphere*, *The Amazing Colossal Man*, *Invasion of the Body Snatchers* and the *Incredible Shrinking Man* were just some of the inspired inventions of that endlessly imaginative decade.

My mother and I never went to these, however. We saw melodramas instead, generally starring people from the lower-middle tiers of the star system – Richard Conte, Lizabeth Scott, Lana Turner, Dan Duryea, Jeff Chandler. I could never understand the appeal of these movies myself. It was all just talk, talk, talk in that gloomy, earnest, accusatory way that people in movies in the Fifties had. The characters nearly always turned away when speaking, so that they appeared inexplicably to be addressing a bookcase or floor lamp rather than the person standing behind them. At some point the music would swell and one of the characters would tell the other (by way of the curtains) that they couldn't take any more of this and were leaving.

'Me, too!' I would quip amiably to my mother and amble off to the men's room for a change of scene. The men's rooms in the downtown theatres were huge, and soothingly lit, and quite splendidly classy. They had good full-length mirrors, so you could practise gunslinger draws, and there were several machines – comb machines, condom machines – you could almost get your arm up. There was a long line of toilet cubicles and they all had those dividers that allowed you to see the feet of people in flanking cubicles, which I never understood, and indeed still don't. It's hard to think of a single circumstance in which seeing the feet of the person next door would be to anyone's advantage. As a kind of signature gesture, I would go into the far left-hand stall and lock the door, then crawl under the divider into the next stall and lock it, and so on down the line until I had locked them all. It always gave me a strange sense of achievement.

Goodness knows what I crawled through in order to accomplish this small feat, but then I *was* enormously stupid. I mean really quite enormously. I remember when I was about six passing almost a whole movie picking some interesting sweet-smelling stuff off the underside of my seat, thinking that it was something to do with the actual manufactured composition of the seat before realizing that it was gum that had been left there by previous users.

I was sick for about two years over what a grotesque and unhygienic activity I had been engaged in and the thought that I had then eaten greasy buttered popcorn and a large packet of Chuckles with the same fingers that had dabbled in other people's abandoned chewings. I had even – oh, yuk! yuk! – licked those fingers, eagerly transferring bucketloads of syphilitic dribblings and uncategorizable swill from their snapped-out Wrigley's and Juicy Fruits to my wholesome mouth and sleek digestive tract. It was only a matter of time – hours at most – before I would sink into a mumbling delirium and in slow, fevered anguish die.

After the movies we always stopped for pie at the Toddle House, a tiny, steamy diner of dancing grease fires, ill-tempered staff and cosy perfection on Grand Avenue.

The Toddle House was little more than a brick hut consisting of a single counter with a few twirly stools, but never has a confined area produced more divine foods or offered a more delicious warmth on a cold night. The pies – flaky of crust, creamy of filling and always generously cut – were heaven on a plate. Normally this was the high point of the evening, but on this night I was distracted and inconsolable. I felt dirty and doomed. I would never have dreamed that worse still could possibly come my way, but in fact it was just about to. As I sat at the counter idly pronging my banana cream pie, feeling sorry for myself and my doomed intestinal tract, I drank from my glass of water and then realized that the old man sitting beside me was drinking from it, too. He was over two hundred years old and had a sort of grey drool at each corner of his mouth. When he put the glass down there were little white masticated bits adrift in the water.

'Akk, akk, akk,' I croaked in quiet horror as I took this in and clutched my throat with both hands. My fork fell noisily to the floor.

'Say, have I bin drinkin' yer water?' he said cheerfully.

'Yes!' I gasped in disbelief, and stared at his plate. 'And you were eating . . . *poached eggs*.'

Poached eggs were the second most obvious food-never-to-share-with-an-underwashed-old-man, exceeded only by cottage cheese – and only barely. As a sort of dribbly by-product of eating, the two were virtually indistinguishable. 'Oh, akk, akk,' I cried and made noises over my plate like a cat struggling to bring up a hairball.

'Well, I sure hope you ain't got no cooties!' he said and slapped me jovially on the back as he got up to pay his bill.

I stared at him dumbfounded. He settled his account, laid a toothpick on his tongue, and sauntered bowlegged out to his pickup truck.

He never made it. As he reached out to open the door, bolts of electricity flew from my wildly dilated eyes and played over his body. He shimmered for an instant, contorted in a brief, silent rictus of agony, and was gone.

It was the birth of ThunderVision. The world had just become a dangerous place for morons.

There are many versions of how the Thunderbolt Kid came to attain his fantastic powers – so many that I am not entirely sure myself, but I believe the first hints that I was not of Planet Earth, but rather from somewhere else (from, as I was later to learn, the Planet Electro in the Galaxy Zizz), lay embedded in the conversations that my parents had. I spent a lot of my childhood listening in on – monitoring really – their chats. They would have immensely long conversations that seemed always to be dancing about on the edge of a curious happy derangement. I remember one day my father came in, quite excitedly, with a word written down on a piece of paper.

'What's this word?' he said to my mother. The word was 'chaise longue'.

'*Shays lounge*,' she said, pronouncing it as all Iowans, perhaps all Americans, did. A chaise longue in those days exclusively signified a type of adjustable patio lounger that had lately become fashionable. They came with a padded cushion that you brought in every night if you thought someone might take it. Our cushion had a coach and four horses galloping across it. It didn't need to come in at night.

'Look again,' urged my father.

'*Shays lounge*,' repeated my mother, not to be bullied.

'No,' he said, 'look at the second word. Look closely.'

She looked. 'Oh,' she said, cottoning on. She tried it again. '*Shays lawn-gway*.'

'Well, it's just "long",' my father said gently, but gave it a gallic purr. '*Shays lohhhnggg*,' he repeated. 'Isn't that something? I must have looked at that word a hundred times and I've never noticed that it wasn't *lounge*.'

'*Lawngg*,' said my mother, marvelling slightly. 'That's going to take some getting used to.'

'It's French,' my father explained.

'Yes, I expect it is,' said my mother. 'I wonder what it means.'

'No idea. Oh, look, there's Bob coming home from work,' my father said, looking out the window. 'I'm going to try it out on him.' So he'd collar Bob in his driveway and they'd have an amazed ten-minute conversation. For the next hour, you would see my father striding up and down the alley, and sometimes into neighbouring streets, with his piece of paper, showing it to neighbours, and they would all have an amazed conversation. Later, Bob would come and ask if he could borrow the piece of paper to show his wife.

It was about this time I began to suspect that I didn't come from this planet and that these people weren't – couldn't be – my biological parents.

Then one day when I was not quite six years old I was in the basement, just poking around, seeing if there was anything sharp or combustible that I hadn't come across before, and hanging behind the furnace I found a woollen jersey of rare fineness. I slipped it on. It was many, many sizes too large for me – the sleeves all but touched the floor if I didn't repeatedly push them back – but it was the handsomest article of attire I had ever seen. It was made of a lustrous oiled wool, deep bottle green in colour, was extremely warm and heavy, rather scratchy and slightly moth-holed but still exceptionally splendid. Across the chest, in a satin material, now much faded, was a golden thunderbolt. Interestingly, no one knew where it came from. My father thought that it might be an old college football or ice hockey jersey, dating from some time before the First World War. But how it got into our house he had no idea. He guessed that the previous owners had hung it there and forgotten it when they moved.

But I knew better. It was, obviously, the Sacred Jersey of Zap, left to me by King Volton, my late natural father, who had brought me to Earth in a silver spaceship in Earth year 1951 (Electron year 21,000,047,002) shortly before our austere but architecturally exuberant planet exploded spectacularly in a billion pieces of rocky debris. He had placed me with this innocuous family in the middle of America and hypnotized them into believing that I was a normal boy, so that I could perpetuate the Electron powers and creed.

This jersey then was the foundation garment of my superpowers. It transformed me. It gave me colossal strength, rippling muscles, X-ray vision, the ability to fly and to walk upside down across ceilings, invisibility on demand, cowboy skills like lassooing and shooting guns out of people's hands from a distance, a good voice for singing around campfires, and curious bluish-black hair with a teasing curl at the crown. It made me, in short, the kind of person that men want to be and women want to be with.

To the jersey I added a range of useful accoutrements from my existing stockpile – Zorro whip and sword, Sky King neckerchief and neckerchief ring (with secret whistle), Robin of Sherwood bow and arrow with quiver, Roy Rogers decorative cowboy vest and bejewelled boots with jingly tin spurs – which further fortified my strength and dazzle. From my belt hung a rattling aluminium army surplus canteen that made everything put into it taste curiously metallic; a compass and official Boy Scout Vitt-L-Kit, providing all the essential implements needed to prepare a square meal in the wilderness and to fight off wildcats, grizzlies and paedophile scoutmasters; a Batman flashlight with signalling attachment (for bouncing messages off clouds); and a rubber bowie knife.

I also sometimes carried an army surplus knapsack containing snack food and spare ammo, but I tended not to use it much as it smelled oddly and permanently of cat urine, and impeded the free flow of the red beach towel that I tied round my neck for flight. For a brief while I wore some underpants over my jeans in the manner of Superman (a sartorial quirk that one struggled to fathom) but this caused such widespread mirth in the Kiddie Corral that I soon gave up the practice.

On my head, according to season, I wore a green felt cowboy hat or Davy Crockett coonskin cap. For aerial work I donned a Johnny Unitas-approved football helmet with sturdy plastic face guard. The whole kit, fully assembled, weighed slightly over seventy pounds. I didn't so much wear it as drag it along with me. When fully dressed I was the Thunderbolt Kid (later Captain Thunderbolt), a name that my father bestowed on me in a moment of chuckling admiration as he unsnagged a caught sword and lifted me up the five wooden steps of our back porch, saving me perhaps ten minutes of heavy climb.

Happily, I didn't need a lot of mobility, for my super-powers were not actually about capturing bad people or doing good for the common man, but primarily about using my X-ray vision to peer beneath the clothes of attractive women and to carbonize and eliminate people – teachers, babysitters, old people who wanted a kiss – who were an impediment to my happiness. All heroes of the day had particular specialities. Superman fought for truth, justice and the American way. Roy Rogers went almost exclusively for Communist agents who were scheming to poison the water supply or otherwise disrupt and insult the American way of life. Zorro tormented an oafish fellow named Sergeant Garcia for obscure but apparently sound reasons. The Lone Ranger fought for law and order in the early west. I killed morons. Still do.

I used to give X-ray vision a lot of thought because I couldn't see how it could work. I mean, if you could see through people's clothing, then surely you would also see through their skin and right into their bodies. You would see blood vessels, pulsing organs, food being digested and pushed through coils of bowel, and much else of a gross and undesirable nature. Even if you could somehow confine your X-rays to rosy epidermis, any body you gazed at wouldn't be in an appealing natural state, but would be compressed and distorted by unseen foundation garments. The breasts, for one thing, would be oddly constrained and hefted, basketed within an unseen bra, rather than relaxed and nicely jiggly. It wouldn't be satisfactory at all – or at least not nearly satisfactory enough. Which is why it was necessary to perfect ThunderVision™, a laser-like gaze that allowed me to strip away undergarments without damaging skin or outer clothing. That ThunderVision, stepped up a grade and focused more intensely, could also be used as a powerful weapon to vaporize irritating people was a pleasing but entirely incidental benefit.

Unlike Superman I had no one to explain to me the basis of my powers. I had to make my own way into the superworld, and find my own role models. This wasn't easy, for although the 1950s was a busy age for heroes, it was a strange one. Nearly all the heroic figures of the day were odd and just a touch unsettling. Most lived with another man, except Roy Rogers, the singing cowboy, who lived with a woman, Dale Evans, who dressed like a man. Batman and Robin looked unquestionably as if they were on their way to a gay mardi gras, and Superman was not a huge amount better. Confusingly, there were actually *two* Supermans. There was the comic-book Superman who had bluish hair, never laughed and didn't take any shit from anybody. And there was the television Superman, who was much more genial and a little bit flabby around the tits, and who actually got wimpier and softer as the years passed.

In similar fashion, the Lone Ranger, who was already not the kind of fellow you would want to share a pup tent with, was made odder still by the fact that the part was played on television by two different actors – by Clayton Moore from 1948 to 1951 and from 1954 to 1957, and by John Hart during the years in between – but the programmes were rerun randomly on local TV, giving the impression that the Lone Ranger not only wore a tiny mask that fooled no one, but changed bodies from time to time. He also had a catchphrase – 'A fiery horse with the speed of light, a cloud of dust and hearty "Hi-ho, Silver": the Lone Ranger' – that made absolutely no sense no matter how you looked at it.

Roy Rogers, my first true hero, was in many ways the most bewildering of all. For one thing, he was strangely anachronistic. He lived in a western town, Mineral City, that seemed comfortably bedded in the nineteenth century. It had wooden sidewalks and hitching posts, the houses used oil lamps, everyone rode horses and carried six-shooters, the marshal dressed like a cowboy and wore a badge – but when people ordered coffee in Dale's café it was brought to them in a glass pot off an electric hob. From time to time modern policemen or FBI men would turn up in cars or even light aeroplanes looking for fugitive Communists and when this happened I can clearly remember thinking, 'What the fuck?' or whatever was the equivalent expression for a five-year-old.

Except for Zorro – who really knew how to make a sword fly – the fights were always brief and bloodless, and never involved hospitalization, much less comas, extensive scarring or death. Mostly they consisted of somebody jumping off a boulder on to somebody passing on a horse, followed by a good deal of speeded-up wrestling. Then the two fighters would stand up and the good guy would knock the bad guy down. Roy and Dale both carried guns – everybody carried guns, including Magnolia, their comical black servant, and Pat Brady, the cook – but never shot to kill. They just shot the pistols out of bad people's hands and then knocked them down with a punch.

The other memorable thing about Roy Rogers – which I particularly recall because my father always remarked on it if he happened to be passing through the room – was that Roy's horse, Trigger, got higher billing than Dale Evans, his wife.

'But then Trigger *is* more talented,' my father would always say.

'And better looking, too!' we would faithfully and in unison rejoin.

Goodness me, but we were happy people in those days.

Chapter 4

THE AGE OF EXCITEMENT

**PRE-DINNER DRINKS WON'T
HARM HEART, STUDY SHOWS**

PHILADELPHIA, PENN. (AP) – A couple of cocktails before dinner, and maybe a third for good measure, won't do your heart any harm. In fact, they may even do some good. A research team at Lankenau Hospital reached this conclusion after a study supported in part by the Heart Association of Southeastern Pennsylvania.

– Des Moines Register, 12 August 1958

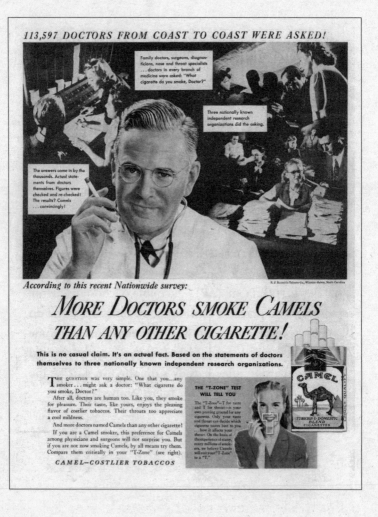

113,597 DOCTORS FROM COAST TO COAST WERE ASKED!

Family doctors, surgeons, diagnosticians, nose and throat specialists ... doctors in every branch of medicine were asked: "What cigarette do you smoke, Doctor?"

Three nationally known independent research organizations did the asking.

The answers come in by the thousands. Actual statements from doctors themselves. Figures were checked and re-checked! The results? Camels ... convincingly!

R. J. Reynolds Tobacco Co., Winston-Salem, North Carolina

According to this recent Nationwide survey:

MORE DOCTORS SMOKE CAMELS
THAN ANY OTHER CIGARETTE!

This is no casual claim. It's an actual fact. Based on the statements of doctors themselves to three nationally known independent research organizations.

THE QUESTION was very simple. One that you...any smoker ... might ask a doctor: "What cigarette do you smoke, Doctor?"

After all, doctors are human too. Like you, they smoke for pleasure. Their taste, like yours, enjoys the pleasing flavor of costlier tobaccos. Their throats too appreciate a cool mildness.

And more doctors named Camels than any other cigarette!

If you are a Camel smoker, this preference for Camels among physicians and surgeons will not surprise you. But if you are not now smoking Camels, by all means try them. Compare them critically in your "T-Zone" (see right).

CAMEL—COSTLIER TOBACCOS

THE "T-ZONE" TEST WILL TELL YOU

The "T-Zone"—T for taste and T for throat—is your own proving ground for any cigarette. Only your taste and throat can decide which cigarette tastes best to you ... how it affects your throat. On the basis of the experience of many, many millions of smokers, we believe Camels will suit your "T-Zone" to a "T".

CAMEL
CHOICE QUALITY
TURKISH & DOMESTIC BLEND
CIGARETTES

I DON'T KNOW HOW they managed it, but the people responsible for the 1950s made a world in which pretty much everything was good for you. Drinks before dinner? The more the better! Smoke? You bet! Cigarettes actually made you healthier, by soothing jangly nerves and sharpening jaded minds, according to advertisements. 'Just what the doctor ordered!' read ads for L&M cigarettes, some of them in the *Journal of the American Medical Association* where cigarette ads were gladly accepted right up to the 1960s. X-rays were so benign that shoe stores installed special machines that used them to measure foot sizes, sending penetrating rays up through the soles of your feet and right out the top of your head. There wasn't a particle of tissue within you that wasn't bathed in their magical glow. No wonder you felt energized and ready for a new pair of Keds when you stepped down.

Happily, we were indestructible. We didn't need seat belts, airbags, smoke detectors, bottled water or the Heimlich manoeuvre. We didn't require child safety caps on our medicines. We didn't need helmets when we rode our bikes or pads for our knees and elbows when we went skating. We knew without written reminding that bleach

105

was not a refreshing drink and that gasoline when exposed to a match had a tendency to combust. We didn't have to worry about what we ate because nearly all foods were good for us: sugar gave us energy, red meat made us strong, ice cream gave us healthy bones, coffee kept us alert and purring productively.

Every week brought exciting news of things becoming better, swifter, more convenient. Nothing was too preposterous to try. 'Mail Is Delivered by Guided Missile', the *Des Moines Register* reported with a clear touch of excitement and pride on the morning of 8 June 1959, after the US Postal Service launched a Regulus I rocket carrying three thousand first-class letters from a submarine in the Atlantic Ocean on to an airbase in Mayport, Florida, one hundred miles away. Soon, the article assured us, rockets loaded with mail would be streaking across the nation's skies. Special delivery letters, one supposed, would be thudding nosecone-first into our back yards practically hourly.

'I believe we will see missile mail developed to a significant degree,' promised Postmaster General Arthur Summerfield at the celebrations that followed. In fact nothing more was heard of missile mail. Perhaps it occurred to someone that incoming rockets might have an unfortunate tendency to miss their targets and crash through the roofs of factories or hospitals, or that they might blow up in flight, or take out passing aircraft, or that every launch would cost tens of thousands of dollars to deliver a payload worth a maximum of $120 at prevailing postal rates.

The fact was that rocket mail was not for one moment a realistic proposition, and that every penny of the million or so dollars spent on the experiment was wasted. No matter. The important thing was knowing that we could send mail by rocket if we *wanted* to. This was an age for dreaming, after all.

Looking back now, it's almost impossible to find anything that wasn't at least a little bit exciting at the time. Even haircuts could give unusual amounts of pleasure. In 1955, my father and brother went to the barber and came back with every hair on their heads standing to attention and sheared off in a perfect horizontal plane. This arresting style was known as a flat-top. They spent most of the rest of the decade looking as if they were prepared in emergencies to provide landing spots for some very small experimental aircraft, or perhaps special delivery messages sent by miniature missile. Never have people looked so ridiculous and so happy at the same time.

There was a certain endearing innocence about the age, too. On 3 April 1956, according to news reports, a Mrs Julia Chase of Hagerstown, Maryland, while on a tour of the White House, slipped away from her tour group and vanished into the heart of the building. For four and a half hours, Mrs Chase, who was described later as 'dishevelled, vague and not quite lucid', wandered through the White House, setting small fires – five in all.

That's how tight security was in those days: a not-quite-lucid woman was able to roam unnoticed through the Executive Mansion for more than half a working day. You can imagine the response if anyone tried anything like that now: the instantaneous alarms, the scrambled Air Force jets, the SWAT teams dropping from panels in the ceiling, the tanks rolling across the lawns, the ninety minutes of sustained gunfire pouring into the target area, the lavish awarding of medals of bravery afterwards, including posthumously to the seventy-six people in Virginia and eastern Maryland killed by friendly fire. In 1956, Mrs Chase, when found, was taken to the staff kitchen, given a cup of tea and released into the custody of her family, and no one ever heard from her again.

Exciting things were happening in the kitchen too. 'A few years ago it took the housewife 5½ hours to prepare daily meals for a family of four,' *Time* magazine reported in a cover article in 1959, which I can guarantee my mother read with great avidity. 'Today she can do it in 90 minutes or less – and still produce meals fit for a king or a finicky husband.' *Time*'s anonymous tipsters went on to list all the fantastic new convenience foods that were just around the corner. Frozen salads. Spray-on mayonnaise. Cheese you could spread with a knife. Liquid instant coffee in a spray can. A complete pizza meal in a tube.

In tones of deepest approbation, the article noted how Charles Greenough Mortimer, chairman of General Foods and a culinary visionary of the first rank, had grown so exasperated with the dullness, the mushiness, the disheartening predictability of conventional vegetables that he had put his best men to work creating 'new' ones in the General Foods laboratories. Mortimer's kitchen wizards had just come up with a product called Rolletes in which they puréed multiple vegetables – peas, carrots and lima beans, for example – and combined the resulting mush into frozen sticks, which the busy homemaker could place on a baking tray and warm up in the oven.

Rolletes went the same way as rocket mail (as indeed did Charles Greenough Mortimer), but huge numbers of other food products won a place in our stomachs and hearts. By the end of the decade the American consumer could choose from nearly one hundred brands of ice cream, five hundred types of breakfast cereals and nearly as many makes of coffee. At the same time, the nation's food factories pumped their products full of delicious dyes and preservatives to heighten and sustain their appeal. By the end of the decade supermarket foods in the United States contained as many as two thousand different chemical additives, including (according to one survey) 'nine emulsifiers, thirty-one stabilizers and thickeners, eighty-five surfactants, seven anti-caking agents, twenty-eight anti-oxidants, and forty-four sequestrants'. Sometimes they contained some food as well, I believe.

Even death was kind of exciting, especially when being safely inflicted on others. In 1951 *Popular Science* magazine asked the nation's ten leading science reporters to forecast the most promising scientific breakthroughs that they expected the next twelve months to bring, and exactly half of them cited refinements to nuclear armaments – several with quite a lot of relish. Arthur J. Snider of the *Chicago Daily News*, for one, noted excitedly that American ground troops could soon be equipped with personal atomic warheads. 'With small atomic artillery capable of firing into concentrations of troops, the ways of tactical warfare are to be revolutionized!' Snider enthused. 'Areas that have in the past been able to withstand weeks and months of siege can now be obliterated in days or hours.' Hooray!

People were charmed and captivated – transfixed, really – by the broiling majesty and unnatural might of atomic bombs. When the military started testing nuclear weapons at a dried lake bed called Frenchman Flat in the Nevada desert outside Las Vegas it became the town's hottest tourist attraction. People came to Las Vegas not to gamble – or at least not exclusively to gamble – but to stand on the desert's edge, feel the ground shake beneath their feet and watch the air before them fill with billowing pillars of smoke and dust. Visitors could stay at the Atomic View Motel, order an Atomic Cocktail ('equal parts vodka, brandy, and champagne, with a splash of sherry') in local cocktail lounges, eat an Atomic Hamburger, get an Atomic Hairdo, watch the annual crowning of Miss Atom Bomb or the nightly rhythmic gyrations of a stripper named Candyce King who called herself 'the Atomic Blast'.

As many as four nuclear detonations a month were conducted in Nevada in the peak years. The mushroom clouds were visible from any parking lot in the city,* but most visitors went to the edge of the blast zone itself, often with picnic lunches, to watch the tests and enjoy the fallout afterwards. And these were big blasts. Some were seen by airline pilots hundreds of miles out over the Pacific Ocean. Radioactive dust often drifted across Las Vegas, leaving a visible coating on every horizontal surface. After some of the early tests, government technicians in white lab coats went through the city running Geiger counters over everything. People lined up to see how radioactive they were. It was all part of the fun. What a joy it was to be indestructible.

Pleasurable as it was to watch nuclear blasts and take on a warm glow of radioactivity, the real joy of the decade – better than flat-tops, rocket mail, spray-on mayonnaise and the atomic bomb combined – was television. It is almost not possible now to appreciate just how welcome TV was.

In 1950, not many private homes in America had televisions. Forty per cent of people still hadn't seen even a single programme. Then I was born and the country

*Though Las Vegas was not in those days the throbbing city we know today. Throughout most of the 1950s it remained a small resort town way out in a baking void. It didn't get its first traffic light until 1952 or its first elevator (in the Riviera Hotel) until 1955.

went crazy (though the two events were not precisely connected). By late 1952, one third of American households – twenty million homes or thereabouts – had purchased TVs. The number would have been even higher except that large parts of rural America still didn't have coverage (or even, often, electricity). In cities, the saturation was much swifter. In May 1953, United Press reported that Boston now had more television sets (780,000) than bathtubs (720,000), and people admitted in an opinion poll that they would rather go hungry than go without their televisions. Many probably did. In the early 1950s, when the average factory worker's after-tax pay was well under $100 a week, a new television cost up to $500.*

TV was so exciting that McGregor, the clothing company, produced a range of clothing in its honour. 'With the spectacular growth of television, millions of Americans are staying indoors,' the company noted in its ads. 'Now, for this revolutionary way of life, McGregor works a sportswear revolution. Whether viewing – or on view – here's sportswear with the new point of view.'

The range was called Videos and to promote it the company produced an illustration, done in the wholesome and meticulous style of a Norman Rockwell painting, showing four athletic-looking young men

*As late as 1959, earnings after tax for a factory worker heading a family of four were $81.03 a week, $73.49 for a single factory worker, though the cost of TVs had fallen significantly.

lounging in a comfortable den before a glowing TV, each sporting a sharp new item from the Videos range – reversible Glen Plaid Visa-Versa Jacket, all-weather Host Tri-Threat Jacket, Durosheen Host Casual Jacket with matching lounge slacks, and, for the one feeling just a touch gay, an Arabian Knights sport shirt in a paisley gabardine, neatly paired with another all-weather jacket. The young men in the illustration look immensely pleased – with the TV, with their outfits, with their good teeth and clear complexions, with everything – and never mind that their clothes are patently designed to be worn out of doors. Perhaps McGregor expected them to stand in neighbours' flowerbeds and watch TV through windows as we did at Mr Kiessler's house. In any case, the McGregor range was not a great success.

People, it turned out, didn't want special clothes for watching television. They wanted special food, and C. A. Swanson and Sons of Omaha came up with the perfect product in 1954: TV dinners (formally TV Brand Dinners), possibly the best bad food ever produced, and I mean that as the sincerest of compliments. TV dinners gave you a whole meal on a compartmentalized aluminium tray. All you had to add was a knife and fork and a dab of butter on the mashed potatoes and you had a complete meal that generally managed (at least in our house) to offer an interesting range of temperature experiences across the compartments, from tepid and soggy (fried chicken) to leap-up-in-astonishment scalding (soup or vegetable) to still partly frozen (mashed potatoes), and all curiously metallic tasting, yet somehow quite satisfying, perhaps simply because it was new and there was nothing else like it. Then some other innovative genius produced special folding trays that you could eat from while watching television, and that was the last time any child – indeed, any male human being – sat at a dining room table voluntarily.

Of course it wasn't TV as we know it now. For one thing, commercials were often built right into the programmes, which gave them an endearing and guileless charm. On *Burns and Allen*, my favourite programme, an announcer named Harry Von Zell would show up halfway through and stroll into George and Gracie's kitchen and do a commercial for Carnation Evaporated Milk ('the milk from contented cows') at the kitchen table while George and Gracie obligingly waited till he was finished to continue that week's amusing story.

Just to make sure that no one forgot that TV was a commercial enterprise, programme titles often generously incorporated the sponsor's name: *The Colgate Comedy Hour*, the *Lux-Schlitz Playhouse*, *The Dinah Shore Chevy Show*, *G.E. Theater*, *Gillette Cavalcade of Sports* and the generously repetitive *Your Kaiser-Fraser Dealer Presents Kaiser-Fraser Adventures in Mystery*. Advertisers dominated every aspect of production. Writers working on shows sponsored by Camel cigarettes were forbidden to show villains smoking cigarettes, to make any mention in any context of fires or arson or anything bad to do with smoke and flames, or to have anyone cough for any reason. When a competitor on the game show *Do You Trust Your Wife?* replied that his wife's astrological sign was Cancer, writes J. Ronald Oakley in the excellent *God's Country: America in the Fifties*, 'the tobacco company sponsoring the show ordered it to be refilmed and the wife's sign changed to Aries.' Even more memorably, for a broadcast of *Judgment at Nuremberg* on a series called Playhouse 90, the sponsor, the American Gas Association, managed to have all references to gas ovens and the gassing of Jews removed from the script.

Only one thing exceeded America's infatuation with television and that was its love of the automobile. Never has a country gone more car-giddy than America did in the 1950s.

When the war ended, there were only thirty million cars on America's roads, roughly the same number as had existed in the 1920s, but then things took off in a big way. Over the next four decades, as a writer for the *New York Times* put it, the country 'paved 42,798 miles of Interstate highway, bought three hundred million cars, and went for a ride'. The number of new cars bought by Americans went from just sixty-nine thousand in 1945 to over five million four years later. By the mid-Fifties Americans were buying eight million new cars a year (this in a nation of approximately forty million households).

They not only wanted to, they *had* to. Under President Eisenhower, America spent three quarters of federal transportation dollars on building highways, and less than 1 per cent on mass transit. If you wanted to get anywhere at all, increasingly you had to do so in your own car. By the middle of the 1950s America was already becoming a two-car nation. As a Chevrolet ad of 1956 exulted: 'The family with two cars gets twice as many chores completed, so there's more leisure to enjoy together!'

And what cars they were. They looked, in the words of one observer, as if they should light up and play. Many boasted features that suggested they might almost get airborne. Pontiacs came with Strato-Streak V8 engines and Strato-Flight Hydra-Matic transmissions. Chryslers offered PowerFlite Range Selector and Torsion-Aire Suspension, while the Chevrolet Bel-Air had a hold-on-to-your-hat feature called Triple-Turbine TurboGlide. In 1958, Ford produced a Lincoln that was over nineteen feet long. By 1961, the American car-buyer had over three hundred and fifty models to choose from.

People were so enamoured of their cars that they more or less tried to live in them. They dined at drive-in restaurants, passed their evenings at drive-in movies, did their banking at drive-in banks, dropped their clothes at drive-in dry cleaners. My father wouldn't have anything to do with any of this. He thought it was somehow unseemly. He wouldn't eat in any restaurant that didn't

116

have booths and a place mat at each setting. (Nor, come to that, would he eat in any place that had anything better than booths and placemats.) So my drive-in experiences came when I went out with Ricky Ramone, who didn't have a dad but whose mom had a red Pontiac Star Chief convertible and *loved* driving fast with the top down and the music way up and going to the A&W drive-in way out by the state fairgrounds on the east side of town, and so I loved her. I'm sure Ricky was conceived in a car, probably between bites at an A&W.

By the end of the decade, America had almost seventy-four million cars on its roads, nearly double the number of ten years before. Los Angeles had more cars than Asia, and General Motors was a bigger economic entity than Belgium, and more exciting, too.

TV and cars went together perfectly. TV showed you a world of alluring things – atomic bombs in Las Vegas, babes on water skis in Cypress Gardens, Florida, Thanksgiving Day parades in New York City – and cars made it possible to get there.

No one understood this better than Walt Disney. When he opened Disneyland on sixty acres of land near the nowhere town of Anaheim, twenty-three miles south of Los Angeles, in 1955, people thought he was out of his mind. Amusement parks were dying in America in the 1950s. They were a refuge of poor people, immigrants, sailors on shore leave and other people of low tone and light pockets. But Disneyland was of course different from the start. First, there was no way to reach it by any form of public transportation, so people of modest means couldn't get there. And if they did somehow contrive to reach the gates, they couldn't afford to get in anyway.*

But Disney's masterstroke was to exploit television for all that it was worth. A year before the park even opened, Disney launched a television series that was essentially a weekly hour-long commercial for Disney enterprises. The programme was actually called *Disneyland* for its first four years and many of the episodes in the series, including the very first, were devoted to celebrating and drumming up interest in that paradise of fantasy and excitement that was swiftly rising from the orange groves at the smoggy end of California.

By the time the park opened, people couldn't wait to get there. Within two years it was attracting four and a half million visitors a year. The average customer, according to *Time* magazine, spent $4.90 on a day out at Disneyland – $2.72 for rides and admission, $2 for food and 18 cents for souvenirs. That seems pretty reasonable to me now – it is awfully hard to believe it wasn't reasonable then – but evidently these were shocking prices. The biggest complaint of Disney customers in the park's first two years, *Time* reported, was the cost.

---

*It says much, I think, that the parking lot at Disneyland, covering one hundred acres, was larger than the park itself, at sixty acres. It could hold 12,175 cars – coincidentally almost exactly the number of orange trees that had been grubbed up during construction.

From our neighbourhood you only went to Disneyland if your father was a brain surgeon or orthodontist. For everyone else, it was too far and too expensive. It was entirely out of the question in our case. My father was a fiend for piling us all in the car and going to distant places, but only if the trips were cheap, educational, and celebrated some forgotten aspect of America's glorious past, generally involving slaughter, uncommon hardship or the delivery of mail at a gallop. Riding in spinning teacups at 15 cents a pop didn't fit into any of that.

The low point of the year in our house came every midwinter when my father retired to his room and vanished into an enormous heap of roadmaps, guidebooks, musty volumes of American history, and brochures from communities surprised and grateful for his interest, to select the destination for our next summer vacation.

'Well, everybody,' he would announce when he emerged after perhaps two evenings' study, 'this year I think we'll tour battlefields of the little-known War of the Filipino Houseboys.' He would fix us with a look that invited cries of rapturous approval.

'Oh, I've never heard of that,' my mom would say cautiously.

'Well, it was actually more of a slaughter than a war,' he would concede. 'It was over in three hours. But it's quite convenient for the National Museum of Agricultural Implements at Haystacks. They have over seven hundred hoes apparently.'

As he spoke he would spread out a map of the western United States, and point to some parched corner of Kansas or the Dakotas that no outsider had ever willingly visited before. We nearly always went west, but never as far as Disneyland and California, or even the Rockies. There were too many Nebraska sodhouses to look at first.

'There's also a steam engine museum at West Windsock,' he would go on happily, and offer a brochure that no one reached for. 'They do a special two-day ticket for families, which looks to be very reasonable. Have you ever seen a steam piano, Billy? No? I'm not surprised. Not many people have!'

The worst thing about going west was that it meant stopping in Omaha on the way home to visit my mother's quizzical relatives. Omaha was an ordeal for everyone, including those we were visiting, so I never understood why we went there, but we always stopped off. It may be that my father was attracted by the idea of free coffee.

My mother grew up remarkably poor, in a tiny house that was really a shack, on the edge of Omaha's vast and famous stockyards. The house had a small back yard, which ended in a sudden cliff, below which, spread out as far as the eye could see (or so it seems in memory), lay the hazy stockyards. Every cow for a thousand miles was brought there to moo hysterically and have a few runny shits before being taken away to become hamburger. You've never smelled such a smell as rose from the stockyards, especially on a hot day, or heard such an unhappy clamour. It was ceaseless and deafening – the sound all but bounced off the clouds – and it made you look twice at all meat products for about a month afterwards.

My mother's father, a good-hearted Irish Catholic named Michael McGuire, had worked the whole of his adult life as a hand in the stockyards on a paltry salary. His wife, my mother's mother, had died when my mother was very small, and he had raised five children more or less single-handed, with my mother and her younger sister Frances doing most of the housework. In her senior year of high school, my mother won a city-wide oration contest which carried as its reward a scholarship to Drake University in Des Moines. There she studied journalism and spent her summers working at the *Register* (where she met my father, a young sportswriter with a broad smile and a weakness for spectacular ties, if old photographs are any guide) and never really came back, something about which I think she always felt a little guilty. Frances eventually went off and became a nun of a timid and twittering disposition. Their father died quite young himself, long before I was born, leaving the house to my mother's three curiously inert brothers, Joey, Johnny and Leo.

It was an astonishment to me even when quite young to think that my mother and her siblings had come from the same genetic stock. I believe she may have felt a little that way herself. My father called her brothers the Three Stooges, though this perhaps suggests a liveliness and joie de vivre, not to mention an entertaining tendency to poke each other in the eyes with forked fingers, that was entirely lacking. They were the three most uninteresting human beings that I have ever met. They had spent their whole lives in this one tiny house, even though they must practically have had to share a bed. I don't know that any of them ever worked or even went outdoors much. The youngest, Leo, had an electric guitar and a small amplifier. If he was asked to play – and he loved nothing better – he would disappear into the bedroom for twenty minutes and emerge, startlingly, in a green sequinned cowboy suit. He knew only two songs, both employing the same chords played in the same order, so fortunately his recitals didn't last long. Johnny spent his whole life sitting at a bare table quietly drinking – he had a fantastic red nose; I mean just fantastic – and Joey had no redeeming qualities at all. When he died, I don't believe anyone was much bothered. I think they may just have rolled his body over the cliff edge. Anyway, when you visited there was nothing to do. I don't recall that they even had a TV. There certainly weren't toys to play with or footballs to kick around. There weren't even enough chairs for everybody to sit down at the same time.

Years later, when Johnny died, my mother discovered he had a common-law wife that he had never told my mother about. I think this wife may actually have been in the closet or under the floorboards or something when we were there. So it is perhaps not surprising that they always seemed kind of keen for us to leave.

Then in 1960, just before my ninth birthday, a really unexpected thing happened. My father announced that we were going to go on a *winter* vacation, over the Christmas holidays, but he wouldn't say where.

It had been an odd fall, but a good one, especially for my dad. My father, you see, was the best baseball writer of his generation – he really was – and in the fall of 1960 I believe he proved it. At a time when most sportswriting was leaden or read as if written by enthusiastic but minimally gifted fourteen-year-olds, he wrote prose that was thoughtful, literate and comparatively sophisticated. 'Neat but not gaudy,' he would always say, with a certain flourish of satisfaction, as he pulled the last sheet out of the type-writer. No one could touch him at writing against a deadline, and on 13 October 1960, at the World Series in Pittsburgh, he put the matter beyond possible dispute.

The Series ended with one of those dramatic moments that baseball seemed to specialize in in those days: Bill Mazeroski of Pittsburgh hit a home run in the ninth inning that snatched triumph from the Yankees and handed it miraculously and unexpectedly to the lowly Pirates. Virtually all the papers in the country reported the news in the same dull, worthy, bewilderingly uninspired tones. Here, for instance, is the opening paragraph of the story that ran on page one of the *New York Times* the next morning:

The Pirates today brought Pittsburgh its first world series baseball championship in thirty-five years when Bill Mazeroski slammed a ninth-inning home run high over the left-field wall of historic Forbes Field.

And here is what people in Iowa read:

The most hallowed piece of property in Pittsburgh baseball history left Forbes Field late Thursday afternoon under a dirty gray sports jacket and with a police escort. That, of course, was home plate, where Bill Mazeroski completed his electrifying home run while Umpire Bill Jackowski, broad back braced and arms spread, held off the mob long enough for Bill to make it legal.

Pittsburgh's steel mills couldn't have made more noise than the crowd in this ancient park did when Mazeroski smashed Yankee Ralph Terry's second pitch of the ninth inning. By the time the ball sailed over the ivy-covered brick wall, the rush from the stands had begun and these sudden madmen threatened to keep Maz from touching the plate with the run that beat the lordly Yankees, 10–9, for the title.

Bear in mind that the story was written not at leisure but amid the din and distraction of a crowded press box in the immediate whooping aftermath of the game. Nor could a single thought or neat phrase (like 'broad back braced and arms spread') have been prepared in advance and casually dropped into the text. Since Mazeroski's home run rudely upended a nation's confident expectations of a victory by 'the lordly Yankees', every sportswriter present had to discard whatever he'd had in mind to say, even one batter earlier, and start afresh. Search as you will, you won't find a better World Series game report on file anywhere, unless it was another of my dad's.*

But I had no idea of this at the time. All I knew was that my father returned home from the Series in unusually high spirits, and revealed his startling plans to take us away on a trip over Christmas to some mysterious locale.

'You wait. You'll like it. You'll see,' was all he would say, to whoever asked. The whole idea of it was unspeakably exciting – we weren't the type of people to do something so rash, so sudden, so *unseasonal* – but unnerving too, for exactly the same reasons. So on the afternoon of 16 December, when Greenwood, my elementary school, dispatched its happy hordes into the snowy streets to begin three glorious weeks of yuletide relaxation (and school holidays in those days, let me say, were of a proper and generous duration), the family Rambler was waiting out front, steaming extravagantly, even keenly, and ready to cut a trail across the snowy prairies. We headed west as

*Of course it's possible I overstate things – this is my father, after all – but if so it is not an entirely private opinion. In 2000, writing in the *Columbia Journalism Review*, Michael Gartner, a former president of NBC News who grew up in Des Moines, wrote that my father, the original Bill Bryson, 'may have been the best baseball writer ever, anywhere'.

usual, crossed the mighty Missouri River at Council Bluffs and made our way past Omaha. Then we just kept on going. We drove for what seemed like (in fact was) days across the endless, stubbly, snow-blown plains. We passed one enticing diversion after another – Pony Express stations, buffalo licks, a pretty big rock – without so much as a sideways glance from my father. My mother began to look faintly worried.

On the third morning, we caught our first sight of the Rockies – the first time in my life I had seen something on the horizon other than a horizon. And still we kept going, up and through the ragged mountains and out the other side. We emerged in California, into warmth and sunshine, and spent a week experiencing its wonders – its mighty groves of redwoods, the lush Imperial Valley, Big Sur, Los Angeles – and the delicious, odd feel of warm sunlight on your face and bare arms in December: a winter without winter.

I had seldom – what am I saying? I had never – seen my father so generous and care-free. At a lunch counter in San Luis Obispo he invited me – *urged* me – to have a large hot fudge sundae, and when I said, 'Dad, are you *sure*?' he said, 'Go on, you only live once' – a sentiment that had never passed his teeth before, certainly not in a commercial setting.

We spent Christmas Day walking on a beach in Santa Monica, and the next day we got in the car and drove south on a snaking freeway through the hazy, warm, endless nowhereness of Los Angeles. At length we parked in an enormous parking lot that was almost comically empty – we were one of only half a dozen cars, all from out of state – and strode a few paces to a grand entrance, where we stood with hands in pockets looking up at a fabulous display of wrought iron.

'Well, Billy, do you know where this is?' my father asked, unnecessarily. There wasn't a child in the world that didn't know these fabled gates.

'It's Disneyland,' I said.

'It certainly is,' he agreed and stared appreciatively at the gates as if they were something he had privately commissioned.

For a minute I wondered if this was all we had come for – to admire the gates – and if in a moment we would get back in the car and drive on to somewhere else. But instead he told us to wait where we were, and strode purposefully to a ticket booth where he conducted a brief but remarkably cheerful transaction. It was the only time in my life that I saw two $20 bills leave my father's wallet simultaneously. As he waited at the window, he gave us a broad smile and a little wave.

'Have I got leukaemia or something?' I asked my mother.

'No, honey,' she replied.

'Has Dad got leukaemia?'

'No, honey, everybody's fine. Your father's just got the Christmas spirit.'

At no point in all my life before or since have I been more astounded, more gratified, more happy than I was for the whole of that day. We had the park practically to ourselves. We did it all – spun gaily in people-sized teacups, climbed aboard flying Dumbos, marvelled at the exciting conveniences in the Monsanto All-Plastic House of the Future in Tomorrowland, enjoyed a submarine ride and riverboat safari, took a rocket to the moon. (The seats actually trembled. 'Whoa!' we all said in delighted alarm.) Disneyland in those days was a considerably less slick and manicured wonder than it would later become, but it was still the finest thing I had ever seen – possibly the finest thing that existed in America at the time. My father was positively enchanted with the place, with its tidiness and wholesomeness and imaginative picture-set charm, and kept asking rhetorically why all the world couldn't be like this. 'But cheaper, of course,' he added, comfortingly returning to character and steering us deftly past a souvenir stand.

The next morning we got in the car and began the thousand-mile trip across desert, mountain and prairie to Des Moines. It was a long drive, but everyone was very happy. At Omaha, we didn't stop – didn't even slow down – but just kept on going. And if there is a better way to conclude a vacation than by not stopping in Omaha, then I don't know it.

Chapter 5
THE PURSUIT OF PLEASURE

In Detroit, Mrs Dorothy Van Dorn, suing for divorce, complained that her husband 1) put all their food in a freezer, 2) kept the freezer locked, 3) made her pay for any food she ate, and 4) charged her the 3% Michigan sales tax.

– *Time* magazine, 10 December 1951

FUN WAS A DIFFERENT KIND OF THING in the 1950s, mostly because there wasn't so much of it. That is not, let me say, a bad thing. Not a great thing perhaps, but not a bad one either. You learned to wait for your pleasures, and to appreciate them when they came.

My most pleasurable experience of these years occurred on a hot day in August 1959 shortly after my mother informed me that she had accepted an invitation on my behalf to go to Lake Ahquabi for the day with Milton Milton and his family. This rash acceptance most assuredly was *not* part of my happiness, believe me, for Milton Milton was the most annoying, the most repellent, the *moistest* drip the world had yet produced, and his parents and sister were even worse. They were noisy, moronically argumentative, told stupid jokes, and ate with their mouths so wide open you could see all the way to their uvulas and some distance beyond. Mr Milton had an Adam's apple the size of a champagne cork and bore as uncanny a resemblance to the Disney character Goofy as was possible without actually being a cartoon dog. His wife was just like him but hairier.

Their idea of a treat was to pass around a plate of Fig

Newtons, the only truly dreadful cookie ever made. They actually yukked when they laughed – an event that gave them a chance to show you just what a well-masticated Fig Newton looks like in its final moments before oblivion (black, sticky, horrible). An hour with the Miltons was like a visit to the second circle of hell. Needless to say, I torched them repeatedly with ThunderVision, but they were strangely ineradicable.

On the one previous occasion on which I had experienced their hospitality, a slumber party at which it turned out I was the only guest, or possibly the only invitee who showed up, Mrs Milton had made me – I'll just repeat that: made me – eat chipped beef on toast, a dish closely modelled on vomit, and then sent us to bed at 8.30 after Milton passed out halfway through *I've Got a Secret*, exhausted after sixteen hours of pretending to be a steam shovel.

So when my mother informed me that she had, in her amiable dementia, committed me to yet another period in their company, my dismay was practically boundless.

'Tell me this isn't happening,' I said and began walking in small, disturbed circles around the carpet. 'Tell me this is just a bad, bad dream.'

'I thought you liked Milton,' said my mother. 'You went to his house for a slumber party.'

'Mom, it was the worst night of my *life*. Don't you remember? Mrs Milton made me eat baked throw-up.

Then she made me share Milton's toothbrush because you forgot to pack one for me.'

'Did I?' said my mother.

I nodded with a kind of strained stoicism. She had packed my sister's toilet bag by mistake. It contained two paper-wrapped tampons and a shower cap, but not my toothbrush or the secret midnight feast that I had been faithfully promised. I spent the rest of the evening playing drums with the tampons on Milton's comatose head.

'I've never been so bored in my life. I *told* you all this before.'

'Did you? I honestly don't recall.'

'Mom, I had to share a toothbrush with Milton Milton after he'd been eating Fig Newtons.'

She received this with a compassionate wince.

'Please don't make me go to Lake Ahquabi with them.'

She considered briefly. 'Well, all right,' she said. 'But I'm afraid you'll have to come with us to visit Sister Gonzaga then.'

Sister Gonzaga was a great-aunt of formidable mien and yet another of the family's many nuns from my mother's side. She was six feet tall and very scary. There was a long-running suspicion in the family that she was actually a man. You always felt that underneath all that starch there was a lot of chest hair. In the summer of 1959, Sister Gonzaga was dying in a local hospital, though not nearly fast enough, if you asked me. Spending an afternoon in Sister Gonzaga's room at the Home for Dying Nuns (I'm not sure that that was its actual name) was possibly the only thing worse than a day out with the Miltons.

So I went to Lake Ahquabi, in a mood of gloomy submission, crammed into the Miltons' ancient, dinky Nash, a car with the comfort and stylish zip of a chest freezer, expecting the worst and receiving it. We got heatedly lost for an hour in the immediate vicinity of the state capitol building – something that was almost impossible for any normal family to do in Des Moines – and when we finally reached Ahquabi spent ninety minutes more, with much additional disputation, unloading the car and setting up a base camp on the shady lawn beside the small artificial beach. Mrs Milton distributed sandwiches, which were made of some kind of pink paste that looked like, and for all I know was, the stuff my grandmother used to secure her dentures to her gums. I went for a little walk with my sandwich and left it with a dog that would have nothing to do with it. Even a procession of ants, I noticed later, had detoured three feet to avoid it.

Having eaten, we had to sit quietly for forty-five minutes before swimming lest we get cramps and die horribly in six inches of water, which was as far in as young males ever ventured on account of perennial rumours that the coffee-coloured depths of Ahquabi harboured vicious snapping turtles that mistook small boys' pizzles for tasty food. Mrs Milton timed this quiet period with an egg timer, and encouraged us to close our eyes and have a little sleep until it was time to swim.

Far out in the lake there was moored a large wooden platform on which stood an improbably high diving board – a kind of wooden Eiffel Tower. It was, I'm sure, the tallest wooden structure in Iowa, if not the Midwest. The platform was so far out from shore that hardly anyone ever visited it. Just occasionally some teenaged daredevils would swim out to have a look around. Sometimes they would climb the many ladders to the high board, and even cautiously creep out on to it, but they always retreated when they saw just how suicidally far the water was below them. No human being had ever been known to jump from it.

So it was quite a surprise when, as the egg timer dinged our liberation, Mr Milton jumped up and began doing neck rolls and arm stretches and announced that he intended to have a dive off the high board. Mr Milton had been a bit of a diving star at Lincoln High School, as he never failed to inform anyone who spent more than three minutes in his company, but that was on a ten-foot board at an indoor pool. Ahquabi was of another order of magnitude altogether. Clearly, he was out of his mind, but Mrs Milton was remarkably untroubled. 'OK, hon,' she replied lazily from beneath a preposterous hat. 'I'll have a Fig Newton for you when you get back.'

Word of the insane intention of the man who looked like Goofy was already spreading along the beach when

Mr Milton jogged into the water and swam with even strokes out to the platform. He was just a tiny, remote stick figure when he got there but even from such a distance the high board seemed to loom hundreds of feet above him – indeed, seemed almost to scrape the clouds. It took him at least twenty minutes to make his way up the zigzag of ladders to the top. Once at the summit, he strode up and down the board, which was enormously long – it had to be to extend beyond the edge of the platform far below – bounced on it experimentally two or three times, then took some deep breaths and finally assumed a position at the fixed end of the board with his arms at his sides. It was clear from his posture and poised manner that he was going to go for it.

By now all the people on the beach and in the water – several hundred altogether – had stopped whatever they were doing and were silently watching. Mr Milton stood for quite a long time, then with a nice touch of theatricality he raised his arms, ran like hell down the long board – imagine an Olympic gymnast sprinting at full tilt towards a distant springboard and you've got something of the spirit of it – took one enormous bounce and launched himself high and outwards in a perfect swan dive. It was a beautiful thing to behold, I must say. He fell with flawless grace for what seemed whole minutes. Such was the beauty of the moment, and the breathless silence of the watching multitudes, that the only sound to be heard across the lake was the faint whistle of his body tearing through the air towards the water far, far below. It may only be my imagination, but he seemed after a time to start to glow red, like an incoming meteor. He was *really* moving.

I don't know what happened – whether he lost his nerve or realized that he was approaching the water at a murderous velocity or what – but about three quarters of the way down he seemed to have second thoughts about the whole business and began suddenly to flail, like someone entangled in bedding in a bad dream, or whose chute hasn't opened. When he was perhaps thirty feet above the water, he gave up on flailing and tried a new tack. He spread his arms and legs wide, in the shape of an X, evidently hoping that exposing a maximum amount of surface area would somehow slow his fall.

It didn't.

He hit the water – *impacted* really is the word for it – at over six hundred miles an hour, with a report so loud that it made birds fly out of trees up to three miles away. At such a speed water effectively becomes a solid. I don't believe Mr Milton penetrated it at all, but just bounced off it about fifteen feet, limbs suddenly very loose, and then lay on top of it, still, like an autumn leaf, spinning gently. He was towed to shore by two passing fishermen in a rowboat, and carried to a grassy area by half a dozen onlookers who carefully set him down on an old blanket. There he spent the rest of the afternoon on his back, arms and legs bent slightly and elevated. Every bit of frontal surface area, from his thinning hairline to his toenails, had a raw, abraded look, as if he had suffered some unimaginable misfortune involving an industrial sander. Occasionally he accepted small sips of water, but otherwise was too traumatized to speak.

Later that same afternoon Milton Junior cut himself with a hatchet that he had been told on no account to touch, so that he ended up in pain and in trouble all at the same time. It was the best day of my life.

Of course, that isn't saying a huge amount when you consider that the previous best day in my life at this point was when Mr Sipkowicz, a teacher we didn't like much, licked a Lincoln Log.

Lincoln Logs were toy wooden logs with which you could build forts, ranchhouses, stockades, bunkhouses, corrals and many other structures of interest and utility to cowboys, according to the imaginative illustrations on the cylindrical box, though in fact the supplied materials were actually just barely enough to make one small rectangular cabin with one door and one window (though you could put the window to the right or left of the door, as you wished).

What Buddy Doberman and I discovered was that if you peed on Lincoln Logs you bleached them white. As a result we created, over a period of weeks, the world's first albino Lincoln Log cabin, which we took to school as part of a project on Abraham Lincoln's early years. Naturally we declined to say how we had made the logs white, prompting pupils and teachers alike to examine them keenly for clues.

'I bet you did it with lemon juice,' said Mr Sipkowicz, who was youthful, brash and odious and had an unfortunate taste for flashy ties, and who for a single semester had the distinction of being Greenwood's only male teacher. Before we could stop him – not that we had any intention or desire to, of course – he shot out a long, reptilian tongue and ran it delicately and experimentally – lingeringly, eye-flutteringly – over the longest log in the back wall, which by chance we had prepared only that morning, so that it was still very slightly moist.

'I can taste lemon, can't I?' he said with a pleased, knowing look.

'Not exactly!!!' we cried and he tried again.

'No, it's lemon,' he insisted. 'I can taste the tartness.' He gave another lick, savouring the flavour with such a deep, concentrated, twitchy intensity that for a moment we thought he had gone into shock and was about to topple over, but it was just his way of relishing the moment. 'Definitely lemon,' he said, brightening, and handed it back to us with great satisfaction all round.

Mr Sipkowicz's unbidden licking gave pleasure, of course, but the real joy of the experience was in knowing that we were the first boys in history to get genuine entertainment out of Lincoln Logs, for Lincoln Logs were inescapably pointless and dull – a characteristic they shared with nearly all other toys of the day.

It would be difficult to say which was the most stupid or disappointing toy of the 1950s since most of them were one or the other, except for those that were both. The one that always leaps to mind for me as most incontestably unsatisfactory was Silly Putty, an oily pink plastic material that did nothing but bounce erratically a dozen or so times before disappearing down a storm drain. (That was actually the best thing about it.) Others, however, might opt for the majestically unamusing Mr Potato Head, a box of plastic parts that allowed children to confirm the fundamental truth that even with ears, limbs and a goofy smile a lifeless tuber is a lifeless tuber.

Also notable for negative ecstasy was Slinky, a coil of metal that could be made to go head over heels down a flight of steps but otherwise did nothing at all, though it did redeem itself slightly from the fact that if you got someone to hold one end – Lumpy Kowalski was always very good for this – and stretched the other end all the way across the street and halfway up a facing slope and then let go, it hit them like a cannon ball. In much the same way, Hula Hoops, those otherwise supremely pointless rings, took on a certain value when used as oversized quoits to ensnare and trip up passing toddlers.

Perhaps nothing says more about the modest range of pleasures of the age than that the most popular candies of my childhood were made of wax. You could choose among wax teeth, wax pop bottles, wax barrels and wax skulls, each filled with a small amount of coloured liquid that tasted very like a small dose of cough syrup. You swallowed this with interest if not exactly gratification, then chewed the wax for the next ten or eleven hours. Now you might think there is something wrong with your concept of pleasure when you find yourself paying real money to chew colourless wax, and you would be right of course. But we did it and enjoyed it because we knew no better. And there was, it must be said, something good, something healthily restrained, about eating a product that had neither flavour nor nutritive value.

You could also get small artificial ice cream cones made of some crumbly chalk-like material, straws containing a gritty sugar so ferociously sour that your whole face would actually be sucked into your mouth like sand collapsing into a hole, root beer barrels, red hot cinnamon balls, liquorice wheels and whips, greasy candy worms, rubbery dense gelatin-like candies that tasted of unfamiliar (and indeed unlikeable) fruits but were good value as it took over three hours to eat each one (and three hours more to pick the gluey remnants out of your molars, sometimes with fillings attached), and jaw-breakers the size and density of billiard balls, which were the best value of all as they would last for up to three months and had different-coloured strata that turned your tongue interesting new shades as you doggedly dissolved away one squamous layer after another.

At Bishop's, where they had a large and highly regarded assortment of penny candies by the cash register, you could also get a comparatively delicious liquorice treat known, with exquisite sensitivity, as nigger babies – though no one actually used that term any more except my grandmother. Occasionally, when visiting from her hometown of Winfield and dining with us at Bishop's, she would slip me a quarter and tell me to go and get some candy for the two of us to share later.

'And don't forget to get some **NIGGER BABIES**!' she would shout, to my intense mortification, across half an acre of crowded dining room, causing a hundred or so diners to look up.

Five minutes later as I returned with the purchase, pressed furtively to outside walls in a vain attempt to escape detection, she would spy me and cry out: 'Oh, there you are, Billy. Did you remember to get some **NIGGER BABIES**? Because I sure do love those . . . **NIGGER BABIES**!'

'Grandma,' I would whisper fiercely, 'you shouldn't say that.'

'Shouldn't say what – **NIGGER BABIES**?'

'Yes. They're called "*liquorice* babies".'

' "Nigger baby" is a bit offensive,' my mom would explain.

'Oh, sorry,' my grandmother would say, marvelling at the delicacy of city people. Then the next time we went to Bishop's, she would say, 'Billy, here's a quarter. Go and get us some of those – whaddaya call 'em – **LIQUORICE NIGGERS**!'

The other place to get penny candies was Grund's, a small grocery store on Ingersoll Avenue. Grund's was one of the last mom and pop grocers left in the city and certainly the last in our neighbourhood. It was run by a doddering couple of adorable minuteness and incalculable antiquity named Mr and Mrs Grund. None of the stock had been renewed, or come to that sold, since about 1929. There were things in there that hadn't been seen in the wider retail world since Gloria Swanson was attractive – Othine skin bleach, Fels-Naptha soap, boxes of Wild Root hair tonic with a photograph of Joe E. Brown on the front. Everything was covered in a thick coating of dust, including Mrs Grund. I believe she may have been dead for some years. Mr Grund, however, was very much alive and delighted when the bell above his door tinklingly sounded the arrival of new customers, even though it was always children and even though they were there for a single nefarious purpose: to steal from his enormous aged stock of penny candies.

This is possibly the most shameful episode of my childhood, but it is one I share with over twelve thousand other former children. Everyone knew you could steal from the Grunds and never be caught. On Saturdays kids turned up from all over the Midwest, some of them arriving in charter buses, if I recall correctly, to stock up for the weekend. Mr Grund was serenely blind to misconduct. You could remove his glasses, undo his bow tie, gently ease him out of his trousers, and he wouldn't suspect a thing. Sometimes we made small purchases, but this was just to get him to turn round and engage his ancient cash register so that a hundred flying hands could dip into his outsized jars and help themselves to more. Some of the bigger kids just took the jars. Still, it has to be said we brightened his day, until we finally put him out of business.

At least candy gave actual pleasure. Most things that were supposed to be fun turned out not to be fun at all. Model making, for instance. Making models was reputed to be hugely enjoyable but it was really just a mysterious ordeal that you had to go through from time to time as part of the boyhood process. The model kits always *looked* fun, to be sure. The illustrations on the boxes portrayed beautifully detailed fighter planes belching red and yellow flames from their wing guns and engaged in lively dog-fights. In the background there was always a stricken Messerschmitt spiralling to earth with a dismayed German in the cockpit, shouting bitter epithets through the wind-screen. You couldn't wait to recreate such lively scenes in three dimensions.

But when you got the kit home and opened the box the contents turned out to be of a uniform leaden grey or olive green, consisting of perhaps sixty thousand tiny parts, some no larger than a proton, all attached in some organic, inseparable way to plastic stalks like swizzle sticks. The tubes of glue by contrast were the size of large pastry tubes. No matter how gently you depressed them they would blurp out a pint or so of a clear viscous goo whose one instinct was to attach itself to some foreign object – a human finger, the living-room drapes, the fur of a passing animal – and become an infinitely long string.

Any attempt to break the string resulted in the creation of more strings. Within moments you would be attached to hundreds of sagging strands, all connected to something that had nothing to do with model aeroplanes or the Second World War. The only thing the glue wouldn't stick to, interestingly, was a piece of plastic model; then it just became a slippery lubricant that allowed any two pieces of model to glide endlessly over each other, never drying. The upshot was that after about forty minutes of intensive but troubled endeavour you and your immediate surroundings were covered in a glistening spider's web of glue at the heart of which was a grey fuselage with one wing on upside down and a pilot accidentally but irremediably attached by his flying cap to the cockpit ceiling. Happily by this point you were so high on the glue that you didn't give a shit about the pilot, the model or anything else.

The really interesting thing about playtime disappointment in the Fifties was that you never saw any of the disappointments coming. This was because the ads were so brilliant. Advertisers have never been so cunning.

They could make any little meretricious piece of crap sound fantastic. Never before or since have commercial blandishments been so silken of tone, so capable of insinuating orgasmic happiness from a few simple materials. Even now in my mind's eye I can see a series of ads in *Boys' Life* from the A. C. Gilbert Company of New Haven, Connecticut, promising the most wholesome joy from their ingenious chemistry sets, microscope kits and world-famous Erector Sets. These last were bolt-together toys from which you could make all manner of engineering marvels – bridges, industrial hoists, fairground rides, motorized robots – from little steel girders and other manly components. These weren't things that you built on table-tops and put in a drawer when you were finished playing. These were items that needed a solid foundation and *lots* of space. I am almost certain that one of the ads showed a boy on a twenty-foot ladder topping out a Ferris wheel on which his younger brother was already enjoying a test ride.

What the ads didn't tell you was that only six people on the planet – A. C. Gilbert's grandsons presumably – had sufficient wealth and roomy enough mansions to enjoy the illustrated sets. I remember my father took one look at the price tag of a giant erection on display in Younkers toy department one Christmas and cried, 'Why, you could practically get a *Buick* for that!' Then he began randomly stopping other male passers-by and soon had a little club of amazed men. So I knew pretty early on that I was never going to get an Erector Set.

Instead I lobbied for a chemistry set, which I had seen in a fetching two-colour double-page spread in *Boys' Life*. According to the ad, this nifty and scientifically advanced kit would allow me to do exciting atomic energy experiments, confound the adult world with invisible writing, become a master of FBI fingerprinting techniques, and make the most satisfyingly enormous stinks. (It didn't actually promise the stinks, but that was implicit in every chemistry set ever sold.)

The set, when opened on Christmas morning, was only about the size of a cigar box – the one portrayed in the magazine had the approximate dimensions of a steamer trunk – but it was ingeniously packed, I must say, with promising stuff: test tubes and a nifty rack in which to set them, a funnel, tweezers, corks, twenty or so little glass pots of colourful chemicals, several of which were promisingly foul smelling, and a plump instruction booklet. Needless to say, I went straight for the atomic energy page, expecting to have a small, private mushroom cloud rising above my workbench by suppertime. In fact, what the instruction book told me, if I recall, was that all materials are made of atoms and that all atoms have energy, so therefore *everything* has atomic energy. Put any two things in a beaker together – any two things at all – give them a shake and, hey presto, you've got an atomic reaction.

All the experiments proved to be more or less like this. The only one that worked even slightly was one of my own devising, which involved mixing together all the chemicals in the set with Babbo cleaning powder, turpentine, some baking soda, two spoonfuls of white pepper, a dab of horseradish of a good age, and a generous splash of Electric-Shave shaving lotion. These when combined instantly expanded about a thousandfold in volume, and ran over the sides of the beaker and on to our brand new kitchen counter, where they began at once to hiss and crinkle and smoke, leaving a pinkish-red welt along the Formica join that would for ever after be a matter of pain and mystification to my father. 'I can't understand it,' he would say, peering along the edge of the counter. 'I must have mixed the adhesive wrong.'

However, the worst toy of the decade, possibly the worst toy ever built, was electric football. Electric football was a game that all boys were compelled to accept as a Christmas present at some point in the 1950s. It consisted of a box with the usual exciting and totally misleading illustrations containing a tinny metal board, about the size of a breakfast tray, painted to look like an American football pitch. This vibrated intensely when switched on, making twenty-two little men move around in a curiously stiff and frantic fashion. It took ages to set up each play because the men were so fiddly and kept falling over, and because you argued continuously with your opponent about what formations were legal and who got to position the last man, since clearly there was an advantage in waiting till the last possible instant and then abruptly moving your running back out to the sidelines where there were no defenders to trouble him. All this always ended in bitter arguments, punctuated by reaching across and knocking over your opponent's favourite players, sometimes repeatedly, with a flicked finger.

It hardly mattered how they were set up because electric football players never went in the direction intended. In practice what happened was that half the players instantly fell over and lay twitching violently as if suffering from some extreme gastric disorder, while the others streamed off in as many different directions as there were upright players, before eventually clumping together in a corner, where they pushed against the unyielding sides like victims of a nightclub fire at a locked exit. The one exception to this was the running back who just trembled in place for five or six minutes, then slowly turned and went on an unopposed glide towards the wrong end zone until knocked over with a finger on the two-yard line by his distressed manager, occasioning more bickering.

At this point you switched off the power, righted all the fallen men, and painstakingly repeated the setting-up process. After three plays like this, one of you would say, 'Hey, do you wanna go and hit Lumpy Kowalski with a stretched Slinky?' and you would push the game out of the way under the bed where it would never be touched again.

The one place where there was real excitement was comic books. This really was the golden age of comics.

Nearly one hundred million of them were being produced every month by the middle of the decade. It is almost impossible to imagine how central a place they played in the lives of the nation's youth – and indeed more than a few beyond youth. A survey of that time revealed that no fewer than 12 per cent of the nation's teachers were devoted readers of comic books. (And that's the ones that admitted it, of course.)

As the Thunderbolt Kid, I read comic books the way doctors read the *New England Journal of Medicine* – to stay abreast of developments in the field. But I was a devoted follower anyway and would have devoured them even without the professional need to keep my supernatural skills honed and productive.

But just as we were getting into comic books, a crisis came. Sales began to falter, pinched between rising production costs and the competition of television. Quite a number of kids now felt that if you could watch Superman and Zorro on TV, why tax yourself with reading words on a page? We in the Kiddie Corral were happy to see such fickle supporters go, frankly, but it was a near-mortal blow for the industry. In two years, the number of comic book titles fell from six hundred and fifty to just two hundred and fifty.

The producers of comic books took some desperate steps to try to rekindle interest. Heroines suddenly became unashamedly sexy. I remember feeling an unexpected but entirely agreeable hormonal warming at the first sight of Asbestos Lady, whose cannonball breasts and powerful loins were barely contained within the wisps of satin fabric with which some artistic genius portrayed her.

There was no space for sentiment in this new age. Captain America's teenaged companion Bucky was dispatched to the hospital with a gunshot wound in one issue and that was the last we ever heard of him. Whether he died or recovered weakly, passing his remaining years in a wheelchair, we didn't know and didn't care. Instead thereafter Captain America was helped by a leggy sylph named Golden Girl, soon augmented by Sun Girl, Lady Lotus, the raven-haired Phantom Lady and other femmes of sleek allure.

Nothing so good could last. Dr Fredric Wertham, a German-born psychiatrist in New York, began an outspoken campaign to rid the world of the baleful influence of comics. In an extremely popular, dismayingly influential book called *Seduction of the Innocent*, he argued that comics promoted violence, torture, criminality, drug-taking and rampant masturbation, though not presumably all at once. Grimly he noted how one boy he interviewed confessed that after reading comic books he 'wanted to be a sex maniac', overlooking that for most boys 'sex', 'mania' and 'want' were words that went together very comfortably with or without comic books.

Wertham saw sex literally in every shadow. He pointed out how in one frame of an action comic the shading on a man's shoulder, when turned at an angle and viewed with an imaginative squint, looked exactly like a woman's pudenda. (In fact it did. There was no arguing the point.) Wertham also announced what most of us knew in our hearts but were reluctant to concede – that many of the superheroes were not fully men in the red-blooded, girl-kissing sense of the term. Batman and Robin in particular he singled out as 'a wish dream of two homosexuals living together'. It was an unanswerable charge. You had only to look at their tights.

Wertham consolidated his fame and influence when he testified before a Senate committee that was looking into the scourge of juvenile delinquency. Just that year Robert Linder, a Baltimore psychologist, had suggested that modern teenagers were suffering from 'a form of collective mental illness' because of rock and roll. Now here was Wertham blaming comics for their sad, zitty failings.

'By 1955,' according to James T. Patterson in the book *Grand Expectations*, 'thirteen states had passed laws regulating the publication, distribution, and sale of comic books.' Alarmed and fearing further regulatory crackdown, the comic book industry abandoned its infatuations with curvy babes, bloody carnage, squint-worthy shadows and everything else that was thrilling. It was a savage blow.

To the dismay of purists, the Kiddie Corral began to fill with anodyne comic books featuring Archie and Jughead or Disney characters like Donald Duck and his nephews Huey, Dewey and Louie, who wore shirts and hats, but nothing at all below the waist, which didn't seem quite right or terribly healthy either. The Kiddie Corral began to attract little girls, who sat chattering away over the latest issues of Little Lulu and Casper the Friendly Ghost as if they were at a tea party. Some perfect fool even put Classic Comic Books in there – the ones that recast famous works of literature in comic book form. These were thrown straight out again, of course.

I vaporized Wertham, needless to say, but it was too late. The damage had been done. Pleasure was going to be harder to get than ever, and the kind we needed most was the hardest of all to get. I refer of course to lust. But that is another story and another chapter.

Chapter 6

SEX AND OTHER DISTRACTIONS

LONDON, ENGLAND (AP) – A high court jury awarded entertainer Liberace 8,000 pounds ($22,400) damages Wednesday in a libel suit against the London Daily Mirror. The jurors decided after 3½ hours of deliberation that a story in 1956 by Mirror journalist William N. Connor implied that the pianist was a homosexual. Among the phrases Liberace cited in his suit was Connor's description of him as 'everything he, she or it can want.' He also described the entertainer as 'fruit-flavored.'

– *Des Moines Tribune*, 18 June 1959

*I dreamed
I went
to work
in my
maidenform*bra*

CHANSONETTE* with famous 'circular-spoke' stitching

Notice <u>two</u> patterns of stitching on the cups of this bra? Circles that uplift and support, spokes that discreetly emphasize your curves. This fine detailing shapes your figure <u>naturally</u>—keeps the bra shapely, even after machine-washing. The triangular cut-out between the cups gives you extra "breathing room" as the lower elastic insert expands. In white or black: A, B, C cups. **2.00**
Other styles: Broadcloth: Cotton, "Dacron" Polyester 3.50; Lace, 3.50; with all-elastic back, 3.00; Contour, 3.00; Full-length, 3.50.
*REG. U.S. PAT. OFF. ©1964 BY **Maidenform, Inc.**, makers of bras, girdles, swimwear, and active sportswear.

IN 1957, THE MOVIE *Peyton Place*, the steamiest motion picture in years, or so the trailers candidly invited us to suppose, was released to a waiting nation and my sister decided that she and I were going to go. Why I was deemed a necessary part of the enterprise I have no idea. Perhaps I provided some sort of alibi. Perhaps the only time she could slip away from the house unnoticed was when she was babysitting me. All I know is that I was told that we were going to walk to the Ingersoll Theater after lunch on Saturday and that I was to tell no one. It was very exciting.

On the way there my sister told me that many of the characters in the movie – probably most of them – would be having sex. My sister at this time was the world's foremost authority on sexual matters, at least as far as I was concerned. Her particular speciality was spotting celebrity homosexuals. Sal Mineo, Anthony Perkins, Sherlock Holmes and Dr Watson, Batman and Robin, Charles Laughton, Randolph Scott, Liberace, of course, and a man in the third row of the Lawrence Welk Orchestra who looked quite normal to me – all were unmasked by her penetrating gaze. She told me Rock Hudson was gay in

1959, long before anyone would have guessed it. She knew that Richard Chamberlain was gay before he did, I believe. She was uncanny.

'Do you know what sex is?' she asked me once we were in the privacy of the woods, walking in single file along the narrow path through the trees. It was a wintry day and I clearly remember that she had on a smart new red woollen coat and a fluffy white hat that tied under her chin. She looked very smart and grown up to me.

'No, I don't believe I do,' I said or words to that effect.

So she told me, in a grave tone and with the kind of careful phrasing that made it clear that this was privileged information, all there was to know about sex, though as she was only eleven at this time her knowledge was perhaps slightly less encyclopaedic than it seemed to me. Anyway, the essence of the business, as I understood it, was that the man put his thing inside her thing, left it there for a bit, and then they had a baby. I remember wondering vaguely what these unspecified things were – his finger in her ear? his hat in her hatbox? Who could say? Anyway, they did this private thing, naked, and the next thing you knew they were parents.

I didn't really care how babies were made, to tell you the truth. I was far more excited that we were on a secret adventure that our parents didn't know about and that we were walking through The Woods – the more or less boundless *Schwarzwald* that lay between Elmwood Drive and Grand Avenue. At six, one ventured into the woods very slightly from time to time, played army a bit within sight of the street and then came out again (usually after Bobby Stimson got poison ivy and burst into tears) with a sense of gladness – of relief, frankly – to be stepping into clear air and sunshine. The woods were unnerving. The air was thicker in there, more stifling, the noises different. You could go into the woods and not come out again. One certainly never considered using them as a thorough-fare. They were far too vast for that. So to be conducted through them by a confident, smart-stepping person, while being given privy information, even if largely meaningless to me, was almost too thrilling for words. I spent most of the long hike admiring the woods' dark majesty and keeping half an eye peeled for gingerbread cottages and wolves.

As if that weren't excitement enough, when we reached Grand Avenue my sister took me down a secret path between two apartment buildings and past the back of Bauder's Drugstore on Ingersoll – it had never occurred to me that Bauder's Drugstore *had* a back – from which we emerged almost opposite the theatre. This was so im-possibly nifty I could hardly stand it. Because Ingersoll was a busy road my sister took my hand and guided us expertly to the other side – another seemingly impossible task. I don't believe I have ever been so proud to be associated with another human being.

At the ticket window, when the ticket lady hesitated, my sister told her that we had a cousin in California who had a role in the movie and that we had promised our mother, a busy woman of some importance ('she's a columnist for the *Register*, you know'), that we would watch the film on her behalf and provide a full report afterwards. As stories go, it was not perhaps the most convincing, but my sister had the face of an angel, a keen manner and that fluffy, innocent hat; it was a combination impossible to disbelieve. So the ticket seller, after a moment's fluttery uncertainty, let us in. I was very proud of my sister for this, too.

After such an adventure, the movie itself was a bit of an anticlimax, especially when my sister told me that we didn't actually have a cousin in the film, or indeed in California. No one got naked and there were no fingers in ears or toes in hatboxes or anything. It was just lots of unhappy people talking to lampshades and curtains. I went off and locked the stalls in the men's room, though as there were only two of them at the Ingersoll even that was a bit disappointing.

By chance, soon afterwards I had an additional experience that shed a little more light on the matter of sex. Coming in from play one Saturday and finding my mother missing from her usual haunts, I decided impulsively to call on my father. He had just returned that day from a long trip to the West Coast – the World Series between the White Sox and Dodgers, as I recall – and we had a lot of catching up to do. So I rushed into his bedroom, expecting to find him unpacking. To my surprise, the shades were drawn and my parents were in bed wrestling under the sheets. More astonishing still, my mother was winning. My father was obviously in some distress. He was making a noise like a small trapped animal.

'What are you doing?' I asked.

'Ah, Billy, your mother is just checking my teeth,' my father replied quickly if not altogether convincingly.

We were all quiet a moment.

'Are you bare under there?' I asked.

'Why, yes we are.'

'Why?'

'*Well*,' my father said as if that was a story that would take some telling, 'we got a bit warm. It's warm work, teeth and gums and so on. Look, Billy, we're nearly finished here. Why don't you go downstairs and we'll be down shortly.'

I believe you are supposed to be traumatized by these things. I can't remember being troubled at all, though it was some years before I let my mother look in my mouth again.

It came as a surprise, when I eventually cottoned on, to realize that my parents had sex – sex between one's parents always seems slightly unbelievable, of course – but also something of a comfort because having sex wasn't easy in the 1950s. Within marriage, with the man on top and woman gritting her teeth, it was just about legal, but almost anything else was forbidden in America in those days. Nearly every state had laws prohibiting any form of sex that was deemed remotely deviant: oral and anal sex of course; homosexuality obviously; even normal, polite sex between consenting but unmarried couples. In Indiana you could be sent to prison for fourteen years for aiding or instigating any person under twenty-one years of age to 'commit masturbation'. The Roman Catholic Archdiocese of the same state declared at about the same time that sex outside marriage was not only sinful, messy and reproductively chancy, but also promoted Communism. Quite how a shag in the haymow helped the relentless march of Marxism was never specified, but it hardly mattered. The point was that once an action was deemed to promote Communism, you knew you were never going to get anywhere near it.

Because lawmakers could not bring themselves to discuss these matters openly, it was often not possible to tell what exactly was being banned. Kansas had (and for all I know still has) a statute vowing to punish, and severely, anyone 'convicted of the detestable and abominable crime against nature committed with mankind or with beast', without indicating even vaguely what a detestable and abominable crime against nature might be. Bulldozing a rainforest? Whipping your mule? There was simply no telling.

Nearly as bad as having sex was thinking about sex. When Lucille Ball on *I Love Lucy* was pregnant for nearly the whole of the 1952–53 season, the show was not allowed to use the word 'pregnant', lest it provoke susceptible viewers to engage in sofa isometrics in the manner of our neighbour Mr Kiessler on St John's Road. Instead, Lucy was described as 'expecting' – a less emotive word apparently. Closer to home, in Des Moines in 1953 police raided Ruthie's Lounge at 1311 Locust Street, and charged the owner, Ruthie Lucille Fontanini, with engaging in an obscene act. It was an act so disturbing that two vice officers and a police captain, Louis Volz, made a special trip to see it – as indeed did most of the men in Des Moines at one time or another, or so it would appear. The act, it turned out, was that Ruthie, with sufficient coaxing from a roomful of happy topers, would balance two glasses on her tightly sweatered chest, fill them with beer and convey them without a spill to an appreciative waiting table.

Ruthie in her prime was a bit of a handful, it would seem. 'She was married sixteen times to nine men,' according to former *Des Moines Register* reporter George Mills in a wonderful book of memoirs, *Looking in Windows*. One of Ruthie's marriages, Mills reported, ended after just sixteen hours when Ruthie woke up to find her new husband going through her purse looking for her safe deposit key. Her custom of using her bosom as a tray would seem a minor talent in an age in which mail was delivered by rocket, but it made her nationally famous. A pair of mountains in Korea were named 'the Ruthies' in her honour and the Hollywood director Cecil B. De Mille visited Ruthie's Lounge twice to watch her in action.

The story has a happy ending. Judge Harry Grund threw the obscenity charges out of court and Ruthie eventually married a nice man named Frank Bisignano and settled down to a quiet life as a housewife. At last report they had been happily married for over thirty years. I'd like to imagine her bringing him ketchup, mustard and other condiments on her chest every evening, but of course I am only guessing.*

For those of us who had an interest in seeing naked women, there were pictures of course in *Playboy* and other manly periodicals of lesser repute, but these were nearly impossible to acquire legally, even if you cycled over to one of the more desperate-looking grocery shacks on the near-east side, lowered your voice two octaves, and swore to God to the impassive clerk that you were born in 1939.

Sometimes in the drugstore if your dad was busy with the pharmacist (and this was the one time I gave sincere thanks for the complex mechanics of isometrics) you could have a rapid shuffle through the pages, but it was a nerve-racking operation as the magazine stand was exposed to view from many distant corners of the store. Moreover, it was right by the entrance and visible from the street through a large plate-glass window, so you were

---

*Ruthie was often described in print as a former stripper. She protested that she had never been a stripper since she had never removed clothes in public. On the other hand, she had often gone onstage without many on.

vulnerable on all fronts. One of your mom's friends could walk past and see you and raise the alarm – there was a police call box on a telephone pole right out front, possibly put there for that purpose – or a pimply stock boy could clamp you on the shoulder from behind and denounce you in a loud voice, or your dad himself could fetch up un-expectedly while you were frantically distracted with trying to locate the pages in which Kim Novak was to be seen relax-ing on a fleecy rug, airing her comely epidermis – so there was practically no pleasure and very little illumination in the exercise. This was an age, don't forget, in which you could be arrested for carrying beer on your bosom or committing an unspecified crime against nature, so what the con-sequences would be if you were caught holding photographs of naked women in a family drugstore was almost inconceivable, but you could be certain they would involve popping flashbulbs, the WHO-TV mobile crime scene unit, banner headlines in the paper, and many thousands of hours of community service.

On the whole therefore you had to make do with underwear spreads in mail order catalogues or ads in glossy magazines, which was desperate to be sure, but at least safely within the law. Maidenform, a maker of brassieres, ran a well-known series of print ads in the 1950s in which women imagined themselves half dressed in public places. 'I dreamed I was in a jewelry store in my Maidenform bra' ran the headline in one, accompanied by a photo showing a woman wearing a hat, skirt, shoes, jewellery and a Maidenform bra – everything, in short, but a blouse – standing at a glass case in Tiffany's or some place like it. There was something deeply – and I expect unhealthily – erotic in these pictures. Unfortunately, Maidenform had an unerring instinct for choosing models of slightly advanced years who were not terribly attractive to begin with and in any case the bras of that period were more like surgical appliances than enticements to fantasy. One despaired at the waste of such a promising erogenous concept.

Despite its shortcomings, the approach was widely copied. Sarong, a manufacturer of girdles so heavy-duty that they looked bulletproof, took a similar line with a series of ads showing women caught by unexpected gusts of wind, revealing their girdles *in situ*, to their own horrified dismay but to the leering delight of all males within fifty yards. I have before me an ad from 1956 showing a woman who has just alighted from a Northwest Airlines flight whose fur coat has inopportunely gusted open (as a result of an extremely localized sirocco occurring somewhere just below and between her legs) to reveal her wearing a Model 124 embroidered nylon marquisette Sarong-brand girdle (available at fine girdlers everywhere for $13.95). But – and here's the thing that has been troubling me since 1956 – the woman is clearly not wearing a skirt or anything else between girdle and coat, raising urgent questions as to how she was dressed when she boarded the plane. Did she fly skirtless the whole way from (let's say for the sake of argument) Tulsa to Minneapolis or did she remove the skirt en route – and why?

Sarong ads had a certain following in my circle – my friend Doug Willoughby was a great admirer – but I always found them strange, illogical and slightly pervy. 'The woman can't have travelled halfway across the country without a skirt on surely,' I would observe repeatedly, even a little heatedly. Willoughby conceded the point without demur, but insisted that that was precisely what made Sarong ads so engaging. Anyway, it's a sad age, you'll agree, when the most titillating thing you can find is a shot of a horrified woman in a half-glimpsed girdle in your mother's magazines.

By chance, we did have the most erotic statue in the nation in Des Moines. It was part of the state's large civil war monument on the capitol grounds. Called 'Iowa', it depicts a seated woman who is holding her bare breasts in her hands, cupped from beneath in a startlingly provocative manner. The pose, we are told, was intended to represent a symbolic offering of nourishment, but really she is inviting every man who goes by to think hard about clambering up and clamping on. We used some-times to ride our bikes there on Saturdays to stare at it from below. 'Erected in 1890' said a plaque on the statue. 'And causing them ever since,' we used to quip. But it was a long way to cycle just to see some copper tits.

The only other option was to spy on people. A boy

named Rocky Koppell, whose family had been transferred to Des Moines from Columbus, lived for a time in an apartment in the basement of the Commodore Hotel and discovered a hole in the wall at the back of his bedroom closet through which he could watch the maid next door dressing and occasionally taking part in an earnest exchange of fluids with one of the janitors. Koppell charged 25 cents to peep through the hole, but lost most of his business when word got around that the maid looked like Adlai Stevenson, but with less hair.

The one place you knew you were never going to see naked female flesh was at the movies. Women undressed in the movies from time to time, of course, but they always stepped behind a screen to do so, or wandered into another room after taking off their earrings and absent-mindedly undoing the top button of their blouse. Even if the camera stayed with the woman, it always shyly dropped its gaze at the critical moment, so that all you saw was a dressing gown falling around the ankles and a foot stepping into the bath. It can't even be described as disappointing because you had no expectations to disappoint. Nudity was just never going to happen.

Those of us who had older brothers knew about a movie called *Mau Mau* that was released in 1955. In its initial manifestation it was a respectable documentary about the Mau Mau uprising in Kenya, soberly narrated by the television newscaster Chet Huntley. But the distributor, a man named Dan Sonney, decided the film wasn't commercial enough. So he hired a local crew of actors and technicians and filmed additional scenes in an orange grove in southern California. These showed topless 'native' women fleeing before men with machetes. These extra scenes he spliced more or less randomly into the existing footage to give the film a little extra pep. The result was a commercial sensation, particularly among boys aged twelve to fifteen. Unfortunately, I was only four in 1955, and so missed out on the only naked celluloid jiggling of the decade.

One year when I was about nine we built a treehouse in the woods – quite a good treehouse, using some first-rate materials appropriated from a construction site on River Oaks Drive – and immediately, and more or less automatically, used it as a place to strip off in front of each other. This was not terribly exciting as the group consisted of about twenty-four little boys and just one girl, Patty Hefferman, who already at the age of seven weighed more than a large piece of earth-moving equipment (she would eventually become known as All-Beef Patty), and was not, with the best will in the world, anyone's idea of Madame Eros. Still, for a couple of Oreo cookies she was willing to be examined from any angle for as long as anyone cared to, which gave her a certain anthropological value.

The only girl in the neighbourhood anybody really wanted to see naked was Mary O'Leary. She was the prettiest child within a million million galaxies, but she wouldn't take her clothes off. She would play in the treehouse happily with us when it was wholesome fun, but the moment things got fruity she would depart by way of the ladder and stand below and tell us with a clenched fury that was nearly tearful that we were gross and loathsome. This made me admire her very much, very much indeed, and often I would depart too (for in truth there was only so much of Patty Hefferman you could take and still eat my mom's cooking) and accompany her to her house, praising her effusively for her virtue and modesty.

'Those guys really are disgusting,' I would say, conveniently overlooking that generally I was one of those guys myself.

Her refusal to take part was in an odd way the most titillating thing about the whole experience. I adored and worshipped Mary O'Leary. I used to sit beside her on her sofa when she watched TV and secretly stare at her face. It was the most perfect thing I had ever seen – so soft, so clean, so ready to smile, so full of rosy light. And there was nothing more perfect and joyous in nature than that face in the micro-instant before she laughed.

In July of that summer, my family went to my grandparents' house for the Fourth of July, where I had the usual dispiriting experience of watching Uncle Dee turn wholesome food into flying stucco. Worse still, my grandparents' television was out of commission and waiting for a new part – the cheerfully moronic local television repairman was unable to see the logic of keeping a supply of spare vacuum tubes in stock, an oversight that earned

him a carbonizing dose of ThunderVision, needless to say
– and so I had to spend the long weekend reading from
my grandparents' modest library, which consisted mostly
of Reader's Digest condensed books, some novels by
Warwick Deeping, and a large cardboard box filled with
Ladies' Home Journals going back to 1942. It was a trying
weekend.

When I returned, Buddy Doberman and Arthur Bergen
were waiting by my house. They barely acknowledged my
parents, so eager were they to get me round the corner to
have a private word. There they breathlessly told me that in
my absence Mary O'Leary had come into the treehouse and
taken her clothes off – every last stitch. She had done so
freely, indeed with a kind of dreamy abandon.

'It was like she was in a trance,' said Bergen fondly.

'A *happy* trance,' added Buddy.

'It was really nice,' said Bergen, his stock of fond
remembrance nowhere near exhausted.

Naturally I refused to believe a word of this. They had
to swear to God a dozen times and hope for their mothers'
deaths on a stack of bibles and much else in a grave vein
before I was prepared to suspend my natural disbelief
even slightly. Above all, they had to describe every
moment of the occasion, something that Bergen was able
to do with remarkable clarity. (He had, as he would boast
in later years, a pornographic memory.)

'Well,' I said, keen as you would expect, 'let's get her
and do it again.'

171

'Oh, no,' Buddy explained. 'She said she wasn't going to do it any more. We had to swear we'd never ask her again. That was the deal.'

'But,' I said, sputtering and appalled, 'that's not fair.'

'The funny thing is,' Bergen went on, 'she said she's been thinking about doing it for a long time, but waited until you weren't there because she didn't want to upset you.'

'Upset me? Upset me? Are you kidding? Upset me? Are you kidding? Are you kidding?'

You can still see the dent in the sidewalk where I beat my head against it for the next fourteen hours. True to her word, Mary O'Leary never came near the treehouse again.

Shortly afterwards, in an inspired moment, I took all the drawers out of my father's closet chest to see what, if anything, was underneath them. I used to strip down his bedroom twice a year, in spring and autumn, when he went to spring training and the World Series, looking for lost cigarettes, stray money and evidence that I was indeed from the planet Electro – perhaps a letter from King Volton or the Electro Congress promising some munificent reward for raising me safely and making sure that my slightest whims were met.

On this occasion, because I had more time than usual on my hands, I took the drawers all the way out to see if anything was behind or beneath them, and so found my father's modest girlie stash, comprising two thin magazines, one called *Dude*, the other *Nugget*. They were extremely cheesy. The women in them looked like Pat Nixon or Mamie Eisenhower – the sort of women you would pay *not* to see naked. I was appalled and astonished, not because my father had men's magazines – this was an entirely welcome development, of course; one to be encouraged by any means possible – but because he had chosen so poorly. It seemed tragically typical of my father that his crippling cheapness extended even to his choice of men's magazines.

Still, they were better than nothing and they did feature unclad women. I took them to the treehouse where they were much prized in the absence of Mary O'Leary. When I returned them to their place ten days or so later, just before he came home from spring training, they were conspicuously well thumbed. Indeed, it was hard not to notice that they had been enjoyed by a wider audience. One was missing its cover and nearly all the pictorials now bore marginal comments and balloon captions, many of a candid nature, in a variety of young hands. Often in the years that followed I wondered what my father made of these spirited emendations, but somehow the moment never seemed right to ask.

Chapter 7

BOOM!

MOBILE, ALA. (AP)– The Alabama Supreme Court yesterday upheld a death sentence imposed on a Negro handyman, Jimmy Wilson, 55, for robbing Mrs. Esteele Barker of $1.95 at her home last year. Mrs. Barker is white.

Although robbery is a capital offense in Alabama, no one has been executed in the state before for a theft of less than $5. A court official suggested that the jury had been influenced by the fact that Mrs. Barker told the jury that Wilson had spoken to her in a disrespectful tone.

A spokesman for the National Association for the Advancement of Colored People called the death sentence 'a sad blot on the nation,' but said the organization is unable to aid the condemned man because it is barred in Alabama.

– *Des Moines Register*, 23 August 1958

AT 7.15 IN THE MORNING local time on 1 November 1952, the United States exploded the first hydrogen bomb in the Eniwetok (or Enewetak or many other variants) Atoll in the Marshall Islands of the South Pacific, though it wasn't really a bomb as it wasn't in any sense portable. Unless an enemy would considerably stand by while we built an 80-ton refrigeration unit to cool large volumes of liquid deuterium and tritium, ran in several miles of cabling and attached scores of electric detonators, we didn't have any way of blowing anyone up with it. Eleven thousand soldiers and civilians were needed to get the device to go off at Eniwetok, so this was hardly the sort of thing you could set up in Red Square without arousing suspicions. Properly, it was a 'thermonuclear device'. Still, it was enormously potent.

Since nothing like this had ever been tried before, nobody knew how big a bang it would make. Even the most conservative estimates, for a blast of five megatons, represented more destructive might than all the firepower used by all sides in the Second World War, and some nuclear physicists thought the explosion might go as high as one hundred megatons – a blast so off the scale that scientists could only guess the chain of consequences. One possibility was that it might ignite all the oxygen in the atmosphere. Still, nothing ventured, nothing annihilated, as the Pentagon might have put it, and on the morning of 1 November somebody lit the fuse and, as I like to picture it, ran like hell.

The blast came in at a little over ten megatons, comparatively manageable but still enough to wipe out a city a thousand times the size of Hiroshima, though of course Earth has no cities that big. A fireball five miles high and four miles across rose above Eniwetok within seconds, billowing into a mushroom cloud that hit the stratospheric ceiling thirty miles above the Earth and spread outwards for over a thousand miles in every direction, disgorging a darkening snowfall of dusty ash as it went, before slowly dissipating. It was the biggest thing of any type ever created by humans. Nine months later the Soviets surprised the western powers by exploding a thermonuclear device of their own. The race to obliterate life was on – and how. Now we truly were become Death, the shatterer of worlds.

So it is perhaps not surprising that as this happened I sat in Des Moines, Iowa, quietly shitting myself. I had little choice. I was ten months old.

What was scary about the growth of the bomb wasn't so much the growth of the bomb as the people in charge of the growth of the bomb. Within weeks of the Eniwetok test the big hats at the Pentagon were actively thinking of ways to put this baby to use. One idea, seriously considered, was to build a device somewhere near the front lines in Korea, induce large numbers of North Korean and Chinese troops to wander over to have a look, and then set it off.

Representative James E. Van Zandt of Pennsylvania, a leading proponent of devastation, promised that soon we would have a device of at least a hundred megatons – the one that might consume all our breathable air. At the same time, Edward Teller, the semi-crazed Hungarian-born physicist who was one of the presiding geniuses behind the development of the H-bomb, was dreaming up exciting peacetime uses for nuclear devices. Teller and his acolytes at the Atomic Energy Commission envisioned using H-bombs to enable massive civil engineering projects on a scale never before conceived – to create open-pit mines where mountains had once stood, to alter the courses of rivers in our favour (ensuring that the Danube, for instance, served only capitalist countries), to blow away irksome impediments to commerce and shipping like the Great Barrier Reef of Australia. Excitedly they reported that just twenty-six bombs placed in a chain across the Isthmus of Panama would excavate a bigger, better Panama Canal more or less at once, and provide a lovely show into the bargain. They even suggested that nuclear devices could be used to alter the Earth's weather by adjusting the amount of dust in the atmosphere, for ever banishing winters from the northern US and sending them permanently to the Soviet Union instead. Almost in passing, Teller proposed that we might use the Moon as a giant target for testing warheads. The blasts would be visible through binoculars from Earth and would provide wholesome entertainment for millions. In short, the creators of the hydrogen bomb wished to wrap the world in unpredictable levels of radiation, obliterate whole ecosystems, despoil the face of the planet, and provoke and antagonize our enemies at every opportunity – and these were their *peacetime* dreams.

But of course the real ambition was to make a gigantically ferocious transportable bomb that we could drop on the heads of Russians and other like-minded irritants whenever it pleased us to do so. That dream became enchanting reality on 1 March 1954, when America detonated fifteen megatons of experimental bang over the Bikini atoll (a place so delightful that we named a lady's swimsuit after it) in the Marshall Islands. The blast exceeded all hopes by a considerable margin. The flash was seen in Okinawa, twenty-six hundred miles away. It threw visible fallout over an area of some seven thousand square miles – all of it drifting in exactly the opposite direction to what was forecast. We were getting good not only at making really huge explosions but at creating consequences that were beyond our capabilities to deal with.

One soldier, based on the island of Kwajalein, described in a letter home how he thought the blast would blow his barracks away. 'All of a sudden the sky lighted up a bright orange and remained that way for what seemed like a couple of minutes . . . We heard very loud rumblings that sounded like thunder. Then the whole barracks began shaking, as if there had been an earthquake. This was followed by a very high wind,' which caused everyone present to grab on to something solid and hold on tight. And this was at a place nearly two hundred miles from the blast site, so goodness knows what the experience was like for those who were even closer – and there were many, among them the unassuming native residents of the nearby island of Rongelap, who had been told to expect a bright flash and a loud bang just before 7 a.m., but had been given no other warnings, no hint that the bang itself might knock down their houses and leave them permanently deafened, and no instructions about dealing with the after-effects. As radioactive ash rained down on them, the puzzled islanders tasted it to see what it was made of – salt, apparently – and brushed it out of their hair.

Within minutes they found they weren't feeling terribly well. No one exposed to the fallout had any appetite for breakfast that morning. Within hours many were severely nauseated and blistering prolifically wherever ash had touched bare skin. Over the next few days, their hair came out in clumps and some started haemorrhaging internally.

Also caught in the fallout were twenty-three puzzled

fishermen on a Japanese boat called, with a touch of irony that escaped no one, the Lucky Dragon. By the time they got back to Japan most of the crewmen were deeply unwell. The haul from their trip was unloaded by other hands and sent to market, where it vanished among the thousands of other catches landed in Japanese ports that day. Unable to tell which fish was contaminated and which not, Japanese consumers shunned fish altogether for weeks, nearly wrecking the industry.

As a nation, the Japanese were none too happy about any of this. In less than ten years they had achieved the unwelcome distinction of being the first victims of both the atom and hydrogen bombs, and naturally they were a touch upset and sought an apology. We declined to oblige. Instead Lewis Strauss, a former shoe salesman who had risen to become chairman of the Atomic Energy Commission (it was that kind of age), responded by suggesting that the Japanese fishermen were in fact Soviet agents.

Increasingly, the United States moved its tests to Nevada, where, as we have seen, people were a good deal more appreciative, though it wasn't just the Marshall Islands and Nevada where we tested. We also set off nuclear bombs on Christmas Island and the Johnston Atoll in the Pacific, above and below water in the South Atlantic Ocean, and in New Mexico, Colorado, Alaska and Hattiesburg, Mississippi (of all places), in the early years of testing. Altogether between 1946 and 1962, the

United States detonated just over a thousand nuclear warheads, including some three hundred in the open air, hurling numberless tons of radioactive dust into the atmosphere. The USSR, China, Britain and France detonated scores more.

It turned out that children, with their trim little bodies and love of milk, were particularly adept at absorbing and holding on to strontium 90 – the chief radioactive product of fallout. Such was our affinity for strontium that in 1958 the average child – which is to say me and thirty million other small people – was carrying ten times more strontium than he had only the year before. We were positively aglow with the stuff.

So the tests were moved underground, but that didn't always work terribly well either. In the summer of 1962, defence officials detonated a hydrogen bomb buried deep beneath the desert of Frenchman Flat, Nevada. The blast was so robust that the land around it rose by some three hundred feet and burst open like a very bad boil, leaving a crater eight hundred feet across. Blast debris went everywhere. 'By four in the afternoon,' the historian Peter Goodchild has written, 'the radioactive dust cloud was so thick in Ely, Nevada, two hundred miles from Ground Zero, that the street lights had to be turned on.' Visible fallout drifted down on six western states and two Canadian provinces – though no one officially acknowledged the fiasco and no public warnings were issued advising people not to touch fresh ash or let their children roll around in it. Indeed, all details of the incident remained secret for two decades until a curious journalist filed suit under the Freedom of Information Act to find out what had happened that day.*

While we waited for the politicians and military to give us an actual Third World War, the comic books were pleased to provide an imaginary one. Monthly offerings with titles like *Atomic War!* and *Atom-Age Combat* began to appear and were avidly sought out by connoisseurs in the Kiddie Corral. Ingeniously, the visionary minds behind these comics took atomic weapons away from the generals and other top brass and put them in the hands of ordinary foot soldiers, allowing them to blow away inexhaustible hordes of advancing Chinese and Russian troops with atomic rockets, atomic cannons, atomic grenades and even atomic rifles loaded with atomic bullets. Atomic bullets! What a concept! The carnage was thrilling. Until Asbestos Lady stole into my life, capturing my young heart and twitchy loins, atomic war comics were the most satisfying form of distraction there was.

*Nuclear testing came to a noisy peak in October 1961 when the Soviets exploded a fifty-megaton device in the Arctic north of the country. (Fifty megatons is equivalent to fifty million tons of TNT – more than three thousand times the force of the Hiroshima blast of 1945, which ultimately killed two hundred thousand people.) The number of nuclear weapons at the peak of the Cold War was sixty-five thousand. Today there are about twenty-seven thousand, all vastly more powerful than those dropped on Japan in 1945, divided between possibly as many as nine countries. More than fifty years after the first atomic tests there, Bikini remains uninhabitable.

Anyway, people had many other far worse things to worry about in the 1950s than nuclear annihilation. They had to worry about polio. They had to worry about keeping up with the Joneses. They had to worry that Negroes might move into the neighbourhood. They had to worry about UFOs. Above all, they had to worry about teenagers. That's right. Teenagers became the number-one fear of American citizens in the 1950s.

There had of course been obnoxious, partly grown human beings with bad complexions since time immemorial, but as a social phenomenon teenagehood was a brand new thing. (The word *teenager* had only been coined in 1941.) So when teens began to appear visibly on the scene, rather like mutant creatures in one of the decade's many outstanding science fiction movies, grown-ups grew uneasy. Teenagers smoked and talked back and petted in the backs of cars. They used disrespectful terms to their elders like 'pops' and 'daddy-o'. They smirked. They drove in endless circuits around any convenient business district. They spent up to fourteen hours a day combing their hair. They listened to rock 'n' roll, a type of charged music clearly designed to get youngsters in the mood to fornicate and smoke hemp. 'We know that many platter-spinners are hop-heads,' wrote the authors of the popular book *USA Confidential*, showing a proud grasp of street patois. 'Many others are Reds, left-wingers or hecklers of social convention.'

Movies like *The Wild One*, *Rebel without a Cause*,

Blackboard Jungle, *High School Confidential*, *Teenage Crime Wave*, *Reform School Girl* and (if I may be allowed a personal favourite) *Teenagers from Outer Space* made it seem that the youth of the nation was everywhere on some kind of dark, disturbed rampage. The *Saturday Evening Post* called juvenile crime 'the Shame of America'. *Time* and *Newsweek* both ran cover stories on the country's new young hoodlums. Under Estes Kefauver the Senate Subcommittee on Juvenile Delinquency launched a series of emotive hearings on the rise of street gangs and associated misbehaviour.

In point of fact, young people had never been so good or so devotedly conservative. More than half of them, according to J. Ronald Oakley in *God's Country: America in the Fifties*, were shown by surveys to believe that masturbation was sinful, that women should stay at home and that the theory of evolution was not to be trusted – views that many of their elders would have warmly applauded. Teenagers also worked hard, and contributed significantly to the nation's well-being with weekend and after-school jobs. By 1955, the typical American teenager had as much disposable income as the average family of four had enjoyed fifteen years earlier. Collectively they were worth $10 billion a year to the national balance sheet. So teenagers weren't bad by any measure. Still it's true, when you look at them now, there's no question that they should have been put down.

* * *

Only one thing came close to matching the fearfulness of teenagers in the 1950s and that was of course Communism. Worrying about Communism was an exhaustingly demanding business in the 1950s. Red danger lurked everywhere – in books and magazines, in government departments, in the teachings of schools, at every place of work. The film industry was especially suspect.

'Large numbers of moving pictures that come out of Hollywood carry the Communist line,' Congressman J. Parnell Thomas of New Jersey, chairman of the House Un-American Activities Committee, gravely intoned to approving nods in 1947, though on reflection no one could actually think of any Hollywood movie that seemed even slightly sympathetic to Marxist thought. Parnell never did specify which movies he had in mind, but then he didn't have much chance to for soon afterwards he was convicted of embezzling large sums from the government in the form of salaries for imaginary employees. He was sentenced to eighteen months in a prison in Connecticut where he had the unexpected pleasure of serving alongside two of the people, Lester Cole and Ring Lardner Junior, whom his committee had put away for refusing to testify.

Not to be outdone, Walt Disney claimed in testimony to HUAC that the cartoonists' guild in Hollywood – run by committed reds and their fellow travellers, he reported – tried to take over his studio during a strike in 1941 with the intention of making Mickey Mouse a Communist. He never produced any evidence either, though he did identify one of his former employees as a Communist because he didn't go to church and had once studied art in Moscow.

It was an especially wonderful time to be a noisy moron. Billy James Hargis, a chubby, kick-ass evangelist from Sapulpa, Oklahoma, warned the nation in weekly sweat-spattered sermons that Communists had insinuated themselves into, and effectively taken over, the Federal Reserve, the Department of Education, the National Council of Churches and nearly every other organization of national standing one could name. His pronouncements were carried on five hundred radio stations and two hundred and fifty television stations and attracted a huge following, as did his many books, which had titles like *Communism: The Total Lie* and *Is the Schoolhouse the Proper Place to Teach Raw Sex?*

Although he had no qualifications (he had flunked out of Ozark Bible College – a rare distinction, one would suppose), Hargis founded several educational establishments, including the Christian Crusade Anti-Communist Youth University. (I would love to have heard the school song.) When asked what was taught at his schools, he replied, 'anti-Communism, anti-Socialism, anti-welfare state, anti-Russia, anti-China, a literal interpretation of the Bible and states' rights'. Hargis eventually came undone when it was revealed that he had had sex with several of his students, male and female alike, during moments of lordly fervour. One couple, according to *The Economist*, made the discovery when they blushingly confessed the misdeed to each other on their wedding night.

At the peak of the Red Scare, thirty-two of the forty-eight states had loyalty oaths of one kind or another. In New York, Oakley notes, it was necessary to swear a loyalty oath to gain a fishing permit. In Indiana loyalty oaths were administered to professional wrestlers. The Communist Control Act of 1954 made it a federal offence to communicate any Communist thoughts by any means, including by semaphore. In Connecticut it became illegal to criticize the government, or to speak ill of the army or the American flag. In Texas you could be sent to prison for twenty years for being a Communist. In Birmingham, Alabama, it was illegal merely to be seen conversing with a Communist.

HUAC issued millions of leaflets entitled 'One Hundred Things You Should Know About Communism', detailing what to look out for in the behaviour of neighbours, friends and family. Billy Graham, the esteemed evangelist, declared that over one thousand decent-sounding American organizations were in fact fronts for Communist enterprises. Rudolf Flesch, author of the best-selling *Why Johnny Can't Read*, insisted that a failure to teach phonics in schools was undermining democracy and paving the way for Communism. Westbrook Pegler, a syndicated columnist, suggested that anyone found to have been a Communist at any time in his life should simply be put to death. Such was the sensitivity, according to David Halberstam, that when General Motors hired a Russian automotive designer named Zora Arkus-Duntov, it described him in press releases, wholly fictitiously, as being 'of Belgian extraction'.

No one exploited the fear to better effect than Joseph R. McCarthy, Republican senator from Wisconsin. In 1950, in a speech in Wheeling, West Virginia, he claimed to have in his pocket a list of two hundred and five Communists working in the State Department. The next day he claimed to have another list with fifty-seven names on it. Over the next four years McCarthy waved many lists, each claiming to show a different number of Communist operatives. In the course of his spirited ramblings he helped to ruin many lives without ever producing a single promised list. Not producing evidence was becoming something of a trend.

Others brought additional prejudices into play. John Rankin, a senior congressman from Mississippi, sagely observed: 'Remember, Communism is Yiddish. I understand that every member of the Politburo around Stalin is either Yiddish or married to one, and that includes Stalin himself.' Against such men, McCarthy looked almost moderate and fairly sane.

Such was the hysteria that it wasn't actually necessary to have done anything wrong to get in trouble. In 1950, three former FBI agents published a book called *Red Channels: The Report of Communist Influence in Radio and*

Television, accusing 151 celebrities – among them Leonard Bernstein, Lee J. Cobb, Burgess Meredith, Orson Welles, Edward G. Robinson and the stripper Gypsy Rose Lee – of various seditious acts. Among the shocking misdeeds of which the performers stood accused were speaking out against religious intolerance, opposing fascism and supporting world peace and the United Nations. None had any connection with the Communist Party or had ever shown any Communist sympathies. Even so, many of them couldn't find work for years afterwards unless (like Edward G. Robinson) they agreed to appear before HUAC as a friendly witness and name names.

Doing anything at all to help Communists became essentially illegal. In 1951, Dr Ernest Chain, a naturalized Briton who had won a Nobel Prize six years earlier for helping to develop penicillin, was barred from entering the United States because he had recently travelled to Czechoslovakia, under the auspices of the World Health Organization, to help start a penicillin plant there. Humanitarian aid was only permissible, it seems, so long as those being saved believed in free markets. Americans likewise found themselves barred from travel. Linus Pauling, who would eventually win two Nobel prizes, was stopped at Idlewild Airport in New York while boarding a plane to Britain, where he was to be honoured by the Royal Society, and had his passport confiscated on the grounds that he had once or twice publicly expressed a liberal thought.

It was even harder for those who were not American by birth. After learning that a Finnish-born citizen named William Heikkilin had in his youth briefly belonged to the Communist Party, Immigration Service employees tracked him down to San Francisco, arrested him on his way home from work, and bundled him on to an aeroplane bound for Europe, with nothing but about a dollar in change and the clothes he was wearing. Not until his plane touched down the following day did officials inform his frantic wife that her husband had been deported. They refused to tell her where he had been sent.

In perhaps the most surreal moment of all, Arthur Miller, the playwright, while facing congressional rebuke and the possibility of prison for refusing to betray friends and theatrical associates, was told that the charges against him would be dropped if he would allow the chairman of HUAC, Francis E. Walter, to be photographed with Miller's famous and dishy wife, Marilyn Monroe. Miller declined.

In 1954, McCarthy finally undid himself. He accused General George Marshall, the man behind the Marshall Plan and a person of unquestioned rectitude, of treason, a charge quickly shown to be preposterous. Then he took on the whole of the United States Army, threatening to expose scores of subversive senior staff that he claimed the Army knowingly shielded within its ranks. In a series of televised hearings lasting thirty-six days in the spring of 1954 and known as the Army–McCarthy hearings, he showed himself to be a bullying, blustering buffoon of the first rank without a shred of evidence against anyone – though in fact he had always shown that. It just took this long for most of the nation to realize it.

Later that year McCarthy was severely censured by the Senate – a signal humiliation. He died three years later in disgrace. But the fact is that had he been just a tiny bit smarter or more likeable, he might well have become President. In any case, McCarthy's downfall didn't slow the assault on Communism. As late as 1959, the New York office of the FBI still had four hundred agents working full time on rooting out Communists in American life, according to Kenneth O'Reilly in *Hoover and the Un-Americans*.

Thanks to our overweening preoccupation with Communism at home and abroad America became the first nation in modern history to build a war economy in peacetime. Defence spending in the Fifties ranged between $40 billion and $53 billion a year – or more than *total* government spending on everything at the dawn of the decade. Altogether the US would lay out $350 billion on defence during the eight years of the Eisenhower Presidency. More than this, 90 per cent of our foreign aid was for military expenditures. We didn't just want to arm ourselves; we wanted to make sure that everybody else was armed, too.

Often, all that was necessary to earn America's enmity, and land yourself in a lot of trouble, was to get in the way of our economic interests. In 1950, Guatemala elected a reformist government – 'the most democratic Guatemala ever had', according to the historian Howard Zinn – under Jacobo Arbenz, an educated landowner of good intentions. Arbenz's election was a blow for the American company United Fruit, which had run Guatemala as a private fiefdom since the nineteenth century. The company owned nearly everything of importance in the country – the ports, the railways, the communications networks, banks, stores and some 550,000 acres of farmland – paid little taxes and could count confidently on the support of a string of repressive dictators.

Some 85 per cent of United Fruit's land was left more or less permanently idle. This kept fruit prices high, but Guatemalans poor. Arbenz, who was the son of Swiss immigrants and something of an idealist, thought this was unfair and decided to remake the country along more democratic lines. He established free elections, ended racial discrimination, encouraged a free press, introduced a forty-hour week, legalized unions and ended government corruption.

Needless to say, most people loved him. In an attempt to reduce poverty, he devised a plan to nationalize, at a fair price, much of the idle farmland – including 1,700 acres of his own – and redistribute it in the form of smallholdings to a hundred thousand landless peasants. To that end Arbenz's government expropriated 400,000 acres of land from United Fruit, and offered as compensation the sum that the company had claimed the land was worth for tax purposes – $1,185,000.

United Fruit now decided the land was worth $16 million actually – a sum the Guatemalan government couldn't afford to pay. When Arbenz turned down United Fruit's demand for the higher level of compensation, the company complained to the United States government, which responded by underwriting a coup.

Arbenz fled his homeland in 1954 and a new, more compliant leader named Carlos Castillo was installed. To help him on his way, the CIA gave him a list of seventy thousand 'questionable individuals' – teachers, doctors, government employees, union organizers, priests – who had supported the reforms in the belief that democracy in Guatemala was a good thing. Thousands of them were never seen again.

And on that sobering note, let us return to Kid World, where the denizens may be small and often immensely stupid, but are at least comparatively civilized.

Chapter 8
SCHOOLDAYS

In Pasadena, California, student Edward Mulrooney was arrested after he tossed a bomb at his psychology teacher's house and left a note that said: 'If you don't want your home bombed or your windows shot out, then grade fairly and put your assignments on the board – or is this asking too much?'

– *Time* magazine, 16 April 1956

GREENWOOD, MY ELEMENTARY SCHOOL, was a wonderful old building, enormous to a small child, like a castle made of brick. Built in 1901, it stood off Grand Avenue at the far end of a street of outstandingly vast and elegant homes. The whole neighbourhood smelled lushly of old money.

Stepping into Greenwood for the first time was both the scariest and most exciting event of the first five years of my life. The front doors appeared to be about twenty times taller than normal doors, and everything inside was built to a similar imposing scale, including the teachers. Everything about it was intimidating and thrilling at once.

It was, I believe, the handsomest elementary school I have ever seen. Nearly everything in it – the cool ceramic water fountains, the polished corridors, the cloakrooms with their ancient, neatly spaced coat hooks, the giant clanking radiators with their intricate embossed patterns like iron veins, the glass-fronted cupboards, everything – had an agreeable creak of solid, classy, utilitarian vener-ability. This was a building made by craftsmen at a time when quality counted, and generations of devoted childhood learning suffused the air. If I hadn't had to spend so much of my time vaporizing teachers I would have adored the place.

Still, I was very fond of the building. One of the glories of life in that ancient lost world of the mid-twentieth century was that facilities designed for kids often were just smaller versions of things in the adult world. You can't imagine how much more splendid this made them. Our Little League baseball field, for instance, was a proper ballpark, with a grandstand and a concession stand and press box, and real dugouts that were, as the name demands, partly subterranean (and never mind that they filled with puddles every time it rained and that the shorter players couldn't see over the edge and so tended to cheer at the wrong moments). When you ran up those three sagging steps and out on to the field you could seriously imagine that you were in Yankee Stadium. Superior infrastructure makes for richer fantasies, believe me. Greenwood contained all that in spades.

It had, for one thing, an auditorium that was just like a real theatre, with a stage with curtains and spotlights and dressing rooms behind. So however bad your school productions were – and ours were always extremely bad, partly because we had no talent and partly because Mrs De Voto, the music teacher, was a bit ancient and often nodded off at the piano – it felt like you were part of a well-ordered professional undertaking (even when you were standing there holding a long note, waiting for Mrs De Voto's chin to touch the keyboard, an event that always jerked her back into action with rousing gusto at exactly the spot where she had left off a minute or two before).

Greenwood also had the world's finest gymnasium. It was upstairs at the back of the school, which gave it a nicely unexpected air. When you opened the door, you expected to find an ordinary classroom and instead you had – hey! whoa! – a gigantic cubic vault of polished wood. It was a space to savour: it had cathedral-sized windows, a ceiling that no ball could ever reach, acres of varnished wood that had been mellowed into a honeyed glow by decades of squeaky sneakers and gentle drops of childish perspiration, and smartly echoing acoustics that made every bouncing ball sound deftly handled and seriously athletic. When the weather was good and we were sent outdoors to play, the route to the playground took us out on to a rickety metal fire escape that was unnervingly but grandly lofty. The view from the summit took in miles of rooftops and sunny countryside reaching practically to Missouri, or so it seemed.

Mostly we played indoors, however, because it was nearly always winter outside. Of course winters in those days, as with all winters of childhood, were much longer, snowier and more frigid than now. We used to get up to eleven feet of snow at a time – we seldom got less, in fact – and weeks of arctic weather so bitter you could pee icicles.

In consequence, they used to keep the school heated to roughly the temperature of the inside of a pottery kiln, so pupils and teachers alike existed in a state of permanent, helpless drowsiness. But at the same time the close warmth made everything deliciously cheery and cosy. Even Lumpy Kowalski's daily plop in his pants smelled oven-baked and kind of strangely lovely. (For six months of the year, his pants actually steamed.) On the other hand, the radiators were so hot that if you carelessly leaned an elbow on them you could leave flesh behind. The most infamous radiator-based activity was of course to pee on a radiator in one of the boys' bathrooms. This created an enormous sour stink that permeated whole wings of the school for days on end and could not be got rid of through any amount of scrubbing or airing. For this reason, anyone caught peeing on a radiator was summarily executed.

The school day was largely taken up with putting on or taking off clothing. It was an exhaustingly tedious process. It took most of the morning to take off your out-door wear and most of the afternoon to get it back on, assuming you could find any of it among the jumbled, shifting heap of garments that carpeted the cloakroom floor to a depth of about three feet. Changing time was always like a scene at a refugee camp, with at least three kids wandering around weeping copiously because they had only one boot or no mittens. Teachers were never to be seen at such moments.

Boots in those days had strange, uncooperative clasps that managed to pinch and lacerate at the same time, producing some really interesting injuries, especially when your hands were numb with cold. The manufacturers might just as well have fashioned the clasps out of razor blades. Because they were so lethal, you ended up leaving the clasps undone, which was more macho but also let in large volumes of snow, so that you spent much of the day in sopping wet socks, which then became three times longer than your feet. In consequence of being constantly damp and hyperthermic, all children had running noses from October to April, which most of them treated as a kind of drip feeder.

Greenwood had no cafeteria, so everybody had to go home for lunch, which meant that we had to dress and undress four times in every school day – six if the teacher was foolish enough to include an outdoor recess at some point. My dear, dim friend Buddy Doberman spent so much of his life changing that he often lost track and would have to ask me whether we were putting hats on or off now. He was always most grateful for guidance.

Among the many thousands of things moms never quite understand – the manliness implicit in grass stains, the satisfaction of a really good burp or other gaseous eructation, the need from time to time to blow into straws as well as suck out of them – winter dressing has always been perhaps the most tragically conspicuous. All moms in the Fifties lived in dread of cold fronts slipping in from Canada, and therefore insisted that their children wear enormous quantities of insulating clothes for at least seven months of the year. This came mostly in the form of underwear – cotton underwear, flannel underwear, long underwear, thermal underwear, quilted underwear, ribbed underwear, underwear with padded shoulders, and possibly more; there was a *lot* of underwear in America in the 1950s – so that you couldn't possibly perish during any of the ten minutes you spent outdoors each day.

What they failed to take into account was that you were so mummified by extra clothing that you had no limb flexion whatever, and if you fell over you would never get up again unless someone helped you, which was not a thing you could count on. Layered underwear also made going to the bathroom an unnerving challenge. The manufacturers did put an angled vent in every item, but these never quite matched up, and anyway if your penis is only the size of a newly budded acorn it's asking a lot to thread it through seven or eight layers of underwear and still maintain a competent handhold. In any visit to the restroom, you would hear at least one cry of anguish from someone who had lost purchase in mid-flow and was now delving frantically for the missing appendage.

Mothers also failed to realize that certain clothes at certain periods of your life would get you beaten up. If, for instance, you wore snowpants beyond the age of six, you got beaten up for it. If you wore a hat with ear flaps or, worse, a chin strap, you could be sure of a beating, or at the very least a couple of scoops of snow down your back. The wimpiest, most foolish thing of all was to wear galoshes. Galoshes were unstylish and ineffective and even the name just sounded stupid and inescapably humiliating. If your mom made you wear galoshes at any point in the year, it was a death sentence. I knew kids who couldn't get prom dates in high school because every girl they asked remembered that they had worn galoshes in third grade.

I was not a popular pupil with the teachers. Only Mrs De Voto liked me, and she liked all the children, largely because she didn't know who any of them were. She wrote 'Billy sings with enthusiasm' on all my report cards, except once or twice when she wrote 'Bobby sings with enthusiasm.' But I excused her for that because she was kind and well meaning and smelled nice.

The other teachers – all women, all spinsters – were large, lumpy, suspicious, frustrated, dictatorial and unkind. They smelled peculiar, too – a mixture of camphor, mentholated mints and the curious belief (which may well have contributed to their spinsterhood) that a generous dusting of powder was as good as a bath. Some of these women had been powdering up for years and believe me it didn't work.

They insisted on knowing strange things, which I found bewildering. If you asked to go to the restroom, they wanted to know whether you intended to do Number 1 or Number 2, a curiosity that didn't strike me as entirely healthy. Besides, these were not terms used in our house. In our house, you either went toity or had a BM

205

(for bowel movement), but mostly you just went 'to the bathroom' and made no public declarations with regard to intent. So I hadn't the faintest idea, the first time I requested permission to go, what the teacher meant when she asked me if I was going to do Number 1 or Number 2.

'Well, I don't know,' I replied frankly and in a clear voice. 'I need to do a big BM. It could be as much as a three or a four.'

I got sent to the cloakroom for that. I got sent to the cloakroom a lot, often for reasons that I didn't entirely understand, but I never really minded. It was a curious punishment, after all, to be put in a place where you were alone with all your classmates' snack foods and personal effects and no one could see what you were getting into. It was also a very good time to get some private reading done.

As a scholar, I made little impact. My very first report card, for the first semester of first grade, had just one comment from the teacher: 'Billy talks in a low tone.' That was it. Nothing about my character or deportment, my sure touch with phonics, my winning smile or can-do attitude, just a terse and enigmatic 'Billy talks in a low tone.' It wasn't even possible to tell whether it was a complaint or just an observation. After the second semester, the report said: 'Billy still talks in a low tone.' All my other report cards – every last one, apart from Mrs De Voto's faithful recording of my enthusiastic noise-making

– had blanks in the comment section. It was as if I wasn't there. In fact, often I wasn't.

Kindergarten, my debut experience at Greenwood, ran for just half a day. You attended either the morning session or the afternoon session. I was assigned to the afternoon group, which was a lucky thing because I didn't get up much before noon in those days. (We were night owls in our house.) One of my very first experiences of kindergarten was arriving for the afternoon, keen to get cracking with the fingerpaints, and being instructed to lie down on a little rug for a nap. Resting was something we had to do a lot of in the Fifties; I presume that it was somehow attached to the belief that it would thwart polio. But as I had only just risen to come to school, it seemed a little eccentric to be lying down again. The next year was even worse because we were expected to turn up at 8.45 in the morning, which was not a time I chose to be active.

My best period was the late evening. I liked to watch the ten o'clock news with Russ Van Dyke, the world's best television newsman (better even than Walter Cronkite), and then *Sea Hunt* starring Lloyd Bridges (some genius at KRNT-TV decided 10.30 at night was a good time to run a show enjoyed by children, which was correct) before settling down with a largish stack of comic books. I was seldom asleep much before midnight, so when my mother called me in the morning, I usually found it inconvenient to rise. So I didn't go to school if I could help it.

I probably wouldn't have gone at all if it hadn't been for mimeograph paper. Of all the tragic losses since the 1950s, mimeograph paper may be the greatest. With its rapturously fragrant, sweetly aromatic pale blue ink, mimeograph paper was literally intoxicating. Two deep drafts of a freshly run-off mimeograph worksheet and I would be the education system's willing slave for up to seven hours. Go to any crack house and ask the people where their dependency problems started and they will tell you, I'm certain, that it was with mimeograph paper in second grade. I used to bound out of bed on a Monday morning because that was the day that fresh mimeographed worksheets were handed out. I draped them over my face and drifted off to a private place where fields were green, everyone went barefoot and the soft trill of pan pipes floated on the air. But most of the rest of the week I either straggled in around mid-morning, or didn't come in at all. I'm afraid the teachers took this personally.

They were never going to like me anyway. There was something about me – my dreaminess and hopeless forgetfulness, my lack of button-cuteness, my permanent default expression of pained dubiousness – that rubbed them the wrong way. They disliked all children, of course, particularly little boys, but of the children they didn't like I believe they especially favoured me. I always did everything wrong. I forgot to bring official forms back on time. I forgot to bring cookies for class parties and Christmas cards and valentines on the appropriate festive days. I always turned up empty-handed for show and tell. I remember once in kindergarten, in a kind of desperation, I just showed my fingers.

If we were going on a school trip, I never remembered to bring a permission note from home, even after being reminded daily for weeks. So on the day of the trip everybody would have to sit moodily on the bus for an interminable period while the principal's secretary tried to track my mother down to get her consent over the phone. But my mother was always out to coffee. The whole fucking women's department was always out to coffee. If they weren't out to coffee, they were out to lunch. It's a miracle they ever got a section out, frankly. The secretary would eventually look at me with a sad smile and we would have to face the fact together that I wasn't going to go.

So the bus would depart without me and I would spend the day in the school library, which I actually didn't mind at all. It's not as if I were missing a trip to the Grand Canyon or Cape Canaveral. This was Des Moines. There were only two places schools went on trips in Des Moines – to the Wonder Bread factory on Second Avenue and University, where you could watch freshly made bread products travelling round an enormous room on conveyor belts under the very light supervision of listless drones in paper hats (and you could be excused for thinking that the purpose of school visits was to give the drones something to stare at), and the museum of the Iowa State Historical Society, the world's quietest and most uneventful building, where you discovered that not a great deal had ever happened in Iowa; nothing at all if you excluded ice ages.

A more regular humiliation was forgetting to bring money for savings stamps. Savings stamps were like savings bonds, but bought a little at a time. You gave the teacher twenty or thirty cents (two dollars if your dad was a lawyer, surgeon or orthodontist) and she gave you a commensurate number of patriotic-looking stamps – one for each dime spent – which you then licked and placed over stamp-sized squares in a savings stamp book. When you had filled a book, you had $10 worth of savings and America was that much closer to licking Communism. I can still see the stamps now: they were a pinkish red with a picture of a minuteman with a three-cornered hat, a musket and a look of resolve. It was a sacred patriotic duty to buy savings stamps.

One day each week – I couldn't tell you which one now; I couldn't tell you which one then – Miss Grumpy or Miss Lesbos or Miss Squat Little Fat Thing would announce that it was time to collect money for US Saving Stamps and every child in the classroom but me would immediately reach into their desk or schoolbag and extract a white envelope containing money and join a line at the teacher's desk. It was a weekly miracle to me that all these other pupils *knew* on which day they were supposed to bring money and then actually *remembered* to do so. That was at least one step of sharpness too many for a Bryson.

One year I had four stamps in my book (two of them pasted in upside down); in all the other years I had zero. My mother and I between us had not remembered once. The Butter boys all had more stamps than I did. Each year the teacher held up my pathetically barren book as an example for all the other pupils of how not to support your country and they would all laugh – that peculiar braying laugh that exists only when children are invited by adults to enjoy themselves at the expense of another child. It is the cruellest laugh in the world.

Despite these self-inflicted hardships, I quite enjoyed school, especially reading. We were taught to read from Dick and Jane books, solid hardbacks bound in a heavy-duty red or blue fabric. They had short sentences in large type and lots of handsome watercolour illustrations featuring a happy, prosperous, good-looking, law-abiding but interestingly strange family. In the Dick and Jane books, Father is always called Father, never Dad or Daddy, and always wears a suit, even for Sunday lunch – even, indeed, to drive to Grandfather and Grandmother's farm for a weekend visit. Mother is always Mother. She is always on top of things, always nicely groomed in a clean frilly apron. The family have no last name. They live in a pretty house with a picket fence on a pleasant street, but they have no radio or TV and their bathroom has no toilet (so no problems deciding between Number 1 and Number 2 in *their* household). The children – Dick, Jane and little

Sally – have only the simplest and most timeless of toys: a ball, a wagon, a kite, a wooden sailboat.

No one ever shouts or bleeds or weeps helplessly. No meals ever burn, no drinks ever spill (or intoxicate). No dust accumulates. The sun always shines. The dog never shits on the lawn. There are no atomic bombs, no Butter boys, no cicada killers. Everyone is at all times clean, healthy, strong, reliable, hard-working, American and white.

Every Dick and Jane story provided some simple but important lesson – respect your parents, share your possessions, be polite, be honest, be helpful, and above all work hard. Work, according to *Growing Up with Dick and Jane*, was the eighteenth new word we learned. I'm amazed it took them that long. Work was what you did in our world.

I was captivated by the Dick and Jane family. They were so wonderfully, fascinatingly different from my own family. I particularly recall one illustration in which all the members of the Dick and Jane family, for entertainment, stand on one leg, hold the other out straight and try to grab a toe on the extended foot without losing balance and falling over. They are having the most wonderful time doing this. I stared and stared at that picture and realized that there were no circumstances, including at gunpoint, in which you could get all the members of my family to try to do that together.

Because our Dick and Jane books at Greenwood were

ten or fifteen years old, they depicted a world that was already gone. The cars were old-fashioned, the buses too. The shops the family frequented were of a type that no longer existed – pet shops with puppies in the window, toy stores with wooden toys, grocers where items were fetched for you by a cheerful man in a white apron. I found everything about this enchanting. There was no dirt or pain in their world. They could even go into Grandfather's hen house to collect eggs and not gag from the stink or become frantically attached to a blob of chicken shit. It was a wonderful world, a perfect world, friendly, hygienic, safe, better than real. There was just one very odd thing about the Dick and Jane books. Whenever any of the characters spoke, they didn't sound like humans.

'Here we are at the farm,' says Father in a typical passage as he bounds from the car (dressed, not incidentally, in a brown suit), then adds a touch robotically: 'Hello, Grandmother. Here we are at the farm.'

'Hello,' responds Grandmother. 'See who is here. It is my family. Look, look! Here is my family.'

'Oh, look! Here we are at the farm,' adds Dick, equally amazed to find himself in a rural setting inhabited by loved ones. He, too, seems to have a kind of mental stuck needle. 'Here we are at the farm,' he goes on. 'Here is Grandfather, too! Here we are at the farm.'

It was like this on every page. Every character talked exactly like people whose brains had been taken away. This troubled me for a long while. One of the great influences of my life in this period was the movie *Invasion of the Body Snatchers*, which I found so convincingly scary that I took it as more or less real, and for about three years I watched my parents extremely closely for tell-tale signs that they had been taken over by alien life forms themselves, before eventually realizing that it would be impossible to tell if they had been; that indeed the first clue that they were turning into pod people would be their becoming *more normal* – and I wondered for a long time if the Dick and Jane family (or actually, for I wasn't completely stupid, the creators of the Dick and Jane family) had been snatched and were now trying to soften us up for a podding of our own. It made sense to me.

I loved the Dick and Jane books so much that I took them home and kept them. (There were stacks of spares in the cloakroom.) I still have them and still look at them from time to time. And I am still looking for a family that would all try to touch their toes together.

Once I had the Dick and Jane books at home and could read them at my leisure, over a bowl of ice cream or while keeping half an eye on the television, I didn't see much need to go to school. So I didn't much go. By second grade I was pretty routinely declining my mother's daily entreaties to rise. It exasperated her to the point of two heavy sighs and some speechless clucking – as close to furious as she ever got – but I realized quite early on that if I just went completely limp and unresponsive and assumed a posture of sacklike uncooperativeness, stirring only very slightly from time to time to mumble that I was really quite seriously unwell and needed rest, she would eventually give up and go away, saying, 'Your dad would be *furious* if he was here now.'

But the thing was he wasn't there. He was in Iowa City or Columbus or San Francisco or Sarasota. He was always somewhere. As a consequence he only learned of these matters twice a year when he was given my report card to review and sign. These always became occasions in which my mother was in as much trouble as I was.

'How can he have 26¼ absences in one semester?' he would say in pained dismay. 'And how, come to that, do you get a quarter of an absence?' He would look at my mother in further pained dismay. 'Do you just send part of him to school sometimes? Do you keep his legs at home?'

My mother would make small fretful noises that didn't really amount to speech.

'I just don't get it,' my father would go on, staring at the report card as if it were a bill for damages unfairly rendered. 'It's gotten beyond a joke. I really think the only solution is a military academy.'

My father had a strange, deep attraction to military academies. The idea of permanent, systematized punishment appealed to a certain dark side of his character. Large numbers of these institutions advertised at the back of the *National Geographic* – why there I don't know – and I would often find those pages bookmarked by him. The ads always showed a worried-looking boy in grey military

215

dress, a rifle many times too big for him at his shoulder, above a message saying something like:

Camp Hardship Military Academy
TEACHING BOYS TO KILL SINCE 1867
We specialize in building character and
eliminating pansy traits.
Write for details at PO Box 1,
Chicken Gizzard, Tenn.

It never came to anything. He would write off for a leaflet – my father was a fiend for leaflets of all types, and catalogues too if they were free – and find out that the fees were as much as for an Austin Healey sports car or a trip to Europe and drop the whole notion, as one might drop a very hot platter. Anyway, I wasn't convinced that military academies were such a bad thing. The idea of being at a place where rifles, bayonets and explosives were at the core of the curriculum had a distinct appeal.

Once a month we had a civil defence drill at school. A siren would sound – a special urgent siren that denoted that this was not a fire drill or storm alert but a nuclear attack by agents of the dark forces of Communism – and everyone would scramble out of their seats and get under their desks with hands folded over heads in the nuclear attack brace position. I must have missed a few of these, for the first time one occurred in my presence I had no idea what was going on and sat fascinated as everyone around me dropped to the floor and parked themselves like little cars under their desks.

'What is this?' I asked Buddy Doberman's butt, for that was the only part of him still visible.

'Atomic bomb attack,' came his voice, slightly muffled. 'But it's OK. It's only a practice, I think.'

I remember being profoundly amazed that anyone would suppose that a little wooden desk would provide a safe haven in the event of an atomic bomb being dropped on Des Moines. But evidently they all took the matter seriously for even the teacher, Miss Squat Little Fat Thing, was inserted under her desk, too – or at least as much of her as she could get under, which was perhaps 40 per cent. Once I realized that no one was watching, I elected not to take part. I already knew how to get under a desk and was confident that this was not a skill that would ever need refreshing. Anyway, what were the chances that the Soviets would bomb Des Moines? I mean, come on.

Some weeks later I aired this point conversationally to my father while we were dining together in the Jefferson Hotel in Iowa City on one of our occasional weekends away, and he responded with a strange chuckle that Omaha, just eighty miles to the west of Des Moines, was the headquarters of Strategic Air Command, from which all American operations would be directed in the event of war. SAC would be hit by everything the Soviets could throw at it, which of course was a great deal. We in

Des Moines would be up to our backsides in fallout within ninety minutes if the wind was blowing to the east, my father told me. 'You'd be dead before bedtime,' he added brightly. 'We all would.'

I don't know which I found more disturbing – that I was at grave risk in a way that I hadn't known about or that my father found the prospect of our annihilation so amusing – but either way it confirmed me in the conviction that nuclear drills were pointless. Life was too short and we'd all be dead anyway. The time would be better spent apologetically but insistently touching Mary O'Leary's budding chest. In any case, I ceased to take part in the drills.

So it was perhaps a little unfortunate that on the morning of my third or fourth drill, Mrs Unnaturally Enormous Bosom, the principal, accompanied by a man in a military uniform from the Iowa Air National Guard, made an inspection tour of the school and espied me sitting alone at my desk reading a comic adventure featuring the Human Torch and that shapely minx Asbestos Lady, surrounded by a roomful of abandoned desks, each sprouting a pair of backward-facing feet and a child's ass.

Boy, was I in trouble. In fact, it was worse than just being straightforwardly in trouble. For one thing, Miss Squat Little Fat Thing was also in trouble for having failed in her supervisory responsibilities and so became deeply, irremediably pissed off at me, and would for ever remain so.

My own disgrace was practically incalculable. I had embarrassed the school. I had embarrassed the principal. I had shamed myself. I had insulted my nation. To be cavalier about nuclear preparedness was only half a step away from treason. I was beyond hope really. Not only did I talk in a low tone, miss lots of school, fail to buy savings stamps and occasionally turn up wearing girlie Capri pants, but clearly I came from a Bolshevik household. I spent more or less the rest of my elementary school career in the cloakroom.

Chapter 9

MAN AT WORK

In Washington, DC, gunman John A. Kendrick testified that he was offered $2,500 to murder Michael Lee, but declined the job because 'when I got done paying taxes out of that, what would I have left?'

– *Time* magazine, 7 January 1953

MARY McGUIRE

Typewriters rampant on field of editorship, deadlines and print . . . D club's own little sweetheart . . . . tiny, crinkly-smiled . . . Mr. McGuire's gift to journalism.

ONCE YOU STRIP OUT all those jobs where people have to look at, touch or otherwise deal with faeces and vomit – sewage workers and hospital bedpan cleaners and so on – being an afternoon newspaper boy in the 1950s and 1960s was possibly the worst job in history. For a start, you had to deliver the afternoon papers six days a week, from Monday through Saturday, and then get up on Sundays before dawn and deliver the Sunday papers too. This was so the regular morning paperboys could enjoy a day off each week. Why they deserved a day of rest and we didn't was a question that appears never to have occurred to anyone except evening newspaper boys.

Anyway, being a seven-day-a-week serf meant that you couldn't go away for an overnight trip or anything fun like that without finding somebody to do the route for you, and that was always infinitely more trouble than it was worth because the stand-in invariably delivered to the wrong houses or forgot to show up or just lost interest halfway through and stuffed the last thirty papers in the big US mailbox at the corner of Thirty-seventh Street and St John's Road, so that you ended up in trouble with the customers, the *Register* and *Tribune*'s circulation manager and the United States postal authorities – and all so that you could have your first day off in one hundred and sixty days. It really wasn't fair at all.

I started as a paperboy when I was eleven. You weren't supposed to be allowed a route until you had passed your twelfth birthday, but my father, keen to see me making my own way in the world and herniated before puberty, pulled some strings at the paper and got me a route early. The route covered the richest neighbourhood in town, around Greenwood School, a district studded with mansions of rambling grandeur.* This sounded like a plum posting, and so it was presented to me by the route manager, Mr McTivity, a man of low ethics and high body odour, but of course mansions have the longest driveways and widest lawns, so it took whole minutes – in some cases, many, many whole minutes – to deliver each paper. And evening papers weighed a ton back then.

Plus I was absent-minded. In those days my hold on the real world was always slight at best, but the combination of long walks, fresh air and lack of distraction left

---

*And these were grand houses. The house known as the Wallace home, an enormous brick heap at the corner of Thirty-seventh Street and John Lynde Road, had been the home of Henry A. Wallace, Vice-President from 1941 to 1945. Among the many worthies who had slept there were two sitting Presidents, Theodore Roosevelt and William Howard Taft, and the world's richest man, John D. Rockefeller. At the time, I knew it only as the home of people who gave very, very small Christmas tips.

me helplessly vulnerable to any stray wisp of fantasy or conjecture that chose to carry me off. I might, for instance, spend a little while thinking about Bizarro World. Bizarro World was a planet that featured in some issues of *Superman* comics. The inhabitants of Bizarro World did everything in reverse – walked backwards, drove backwards, switched televisions off when they wanted to watch and on when they didn't, drove through red lights but stopped at green ones, and so on. Bizarro World bothered me enormously because it was so impossibly inconsistent. The people didn't actually speak backwards, but just talked in a kind of primitive cave man 'me no like him' type of English, which was not the same thing at all. Anyway, living backwards simply couldn't be made to work. At the gas station they would have to take fuel out of their cars rather than put it in, so how would they make their cars go? Eating would mean sucking poo up through their anus, sending it through the body and ejecting it in mouth-sized lumps on to forks and spoons. It wouldn't be satisfactory at all.

When I had exhausted that topic, I would generally devote a good stretch of time to 'what if' questions – what I would do if I could make myself invisible (go to Mary O'Leary's house about bath time), or if time stopped and I was the only thing on earth left moving (take a lot of money from a bank and then go to Mary O'Leary's house) or if I could hypnotize everyone in the world (ditto) or found a magic lamp and was granted two wishes (ditto) or anything at all really. All fantasies led ultimately to Mary O'Leary.

Then I might move on to imponderables. How could we be sure that we all saw the same colours? Maybe what I see as green you see as blue. Who could actually say? And when scientists say that dogs and cats are colour-blind (or not – I could never remember which it was), how do they *know*? What dog is going to tell them? How do migrating birds know which one to follow? What if the lead bird just wants to be alone? And when you see two ants going in opposite directions pause to check each other out, what information exactly are they exchanging? – 'Hey, nice feelers!', 'Don't panic, but that kid that's watching us has got matches and lighter fluid' – and how do they know to do whatever they are doing? *Something* is telling them to go off and bring home a leaf or a granule of sand – but who and how?

And then suddenly I would realize that I couldn't remember, hadn't actually consciously experienced, any of the last forty-seven properties I had visited, and didn't know if I had left a paper or just walked up to the door, stood for a moment like an underfunctioning automaton and turned round and walked away again.

It is not easy to describe the sense of self-disappointment that comes with reaching the end of your route and finding that there are sixteen undelivered papers in your bag and you don't have the least idea – not the least idea – to whom they should have gone. I spent much of my prepubescent years first walking an enormous newspaper route, then revisiting large parts of it. Sometimes twice.

As if delivering papers seven days a week weren't enough, you also had to collect the subscription money. So at least three evenings a week, when you might instead have had your feet up and been watching *Combat* or *The Outer Limits*, you had to turn out again and try to coax some money out of your ungrateful customers. That was easily the worst part. And the worst part of the worst part was collecting from Mrs Vandermeister.

Mrs Vandermeister was seven hundred years old, possibly eight hundred, and permanently attached to an aluminium walker. She was stooped, very small, forgetful, glacially slow, interestingly malodorous, practically deaf. She emerged from her house once a day to drive to the supermarket, in a car about the size of an aircraft carrier. It took her two hours to get out of her house and into the car and then another two hours to get the car out of the driveway and up the alley. Partly this was because Mrs Vandermeister could never find a gear she liked and partly because when shunting she never moved forward or backward more than a quarter of an inch at a time, and seemed only barely in touch with the necessity of occasionally turning the wheel. Everyone on the alley knew not to try to go anywhere between 10 a.m. and noon because Mrs Vandermeister would be getting her car out.

Once on the open road, Mrs Vandermeister was

famous over a much wider area. Though her trip to Dahl's was only about three quarters of a mile, her progress created scenes reminiscent of the streets of Pamplona when the bulls are running. Motorists and pedestrians alike fled in terror before her. And it was, it must be said, an unnerving sight when Mrs Vandermeister's car came towards you down the street. For a start, it looked as if it was driverless, such was her exceeding diminutiveness, and indeed it drove as if driverless, for it was seldom entirely on the road, particularly when bumping round corners. Generally there were sparks coming off the under-carriage from some substantial object – a motorcycle, a garbage can, her own walking frame – that she had collected en route and was now taking with her wherever she went.

Getting money from Mrs Vandermeister was a perennial nightmare. Her front door had a small window in it that provided a clear view down her hallway to her living room. If you rang the doorbell at fifteen-second intervals for an hour and ten minutes, you knew that eventually she would realize someone was at the door – 'Now who the heck is *that*!' she would shout to herself – and begin the evening-long process of getting from her chair to the front door, twenty-five feet away, bumping and shoving her walker before her. After about twenty minutes, she would reach the hallway and start coming towards the door at about the speed that ice melts. Sometimes she would forget where she was going and start to detour into the kitchen or bathroom, and you would have to ring the doorbell like fury to get her back on course. When eventually she came to the door, you would have an extra half-hour of convincing her that you were not a murderer.

'I'm the paperboy, Mrs Vandermeister!' you would shout at her through the little glass pane.

'Billy Bryson's my paperboy!' she would shout back at the doorknob.

'I *am* Billy Bryson! Look at me through the window, Mrs Vandermeister! Look up here! You can see me if you look up here, Mrs Vandermeister!'

'Billy Bryson lives three doors down!' Mrs Vandermeister would shout. 'You've come to the wrong house! I don't know why you've come here!'

'Mrs Vandermeister, I'm collecting for the paper! You owe me three dollars and sixty cents!'

When finally you persuaded her to haul open the door, she was always surprised to find you there – 'Oh, Billy, you gave me a start!' she'd say – and then there would be another small eternity while she went off, shuffling and wobbling and humming the Alzheimer theme tune, to find her purse, a half-hour more while she came back to ask how much again, another forgetful detour to toilet or kitchen, and finally the announcement that she didn't have that much cash and I'd have to call again on a future occasion.

'You shouldn't leave it so long,' she'd shout. 'It's only

supposed to be a dollar twenty every two weeks. You tell Billy when you see him.'

At least Mrs Vandermeister had the excuse of being ancient and demented. What really maddened was being sent away by normal people, usually because they couldn't be bothered to get their purses out. The richer the people were the more likely they were to send you away – always with a fey can-you-ever-forgive-me smile and an apology.

'No, it's all right, lady. I'm very happy to hike a mile and a quarter here through three feet of snow on the coldest night of the year and leave empty-handed because you've got some muffins in the fucking oven and your nails are drying. No problem!'

Of course I never said anything like that, but I did start levying fines. I would add fifty or sixty cents to rich people's bills and tell them that it was because the month started on a Wednesday so there was an extra half-week to account for. You could show them on their kitchen calendar how there were an extra few days at the beginning or end of the month. This always worked, especially with men if they'd had a cocktail or two, and they always had.

'Son of a gun,' they'd say, shaking their head in wonder, while you pocketed their extra money.

'You know, maybe your boss isn't paying you the right amount each month,' I would sometimes pleasantly add.

'Yeah – hey, *yeah*,' they'd say and look really unsettled.

The other danger of rich people was their dogs. Poor people in my experience have mean dogs and know it. Rich people have mean dogs and refuse to believe it. There were thousands of dogs in those days, too, inhabiting every property – big dogs, grumpy dogs, stupid dogs, tiny nippy irritating little dogs that you positively ached to turn into a kind of living hacky-sack, dogs that wanted to smell you, dogs that wanted to sit on you, dogs that barked at everything that moved. And then there was Dewey. Dewey was a black Labrador, owned by a family on Terrace Drive called the Haldemans. Dewey was about the size of a black bear and hated me. With any other human being he was just a big slobbery bundle of softness. But Dewey wanted me dead for reasons he declined to make clear and I don't believe actually knew himself. He just took against me. The Haldemans laughingly dismissed the idea that Dewey had a mean streak and serenely ignored any suggestions that he ought to be kept tied up, as the law actually demanded. They were Republicans – Nixon Republicans – and so didn't subscribe to the notion that laws are supposed to apply to all people equally.

I particularly dreaded Sunday mornings when it was dark because Dewey was black and invisible, apart from his teeth, and it was just him and me in a sleeping world. Dewey slept wherever unconsciousness overtook him – sometimes on the front porch, sometimes on the back

231

porch, sometimes in an old kennel by the garage, sometimes on the path, but always outside – so he was always
there, and always no more than a millimetre away from
wakefulness and attack. It took me ages to creep, breath
held, up the Haldemans' front walk and up the five wide
wooden creak-ready steps of their front porch and very,
very gently set the paper down on the mat, knowing that
at the moment of contact I would hear from some place
close by but unseen a low, dark, threatening growl that
would continue until I had withdrawn with respectful
backward bows. Occasionally – just often enough to leave
me permanently scarred and unnerved – Dewey would
lunge, barking viciously, and I had to fly across the yard
whimpering, hands held protectively over my butt, leap
on my bike and pedal wildly away, crashing into fire
hydrants and lamp posts and generally sustaining far
worse injuries than if I had just let Dewey hold me down
and gnaw on me a bit.

The whole business was terrible beyond words. The
only aspect worse than suffering an attack was waiting for
the next one. The lone redeeming feature of life with
Dewey was the rush of relief when it was all over, of
knowing that I wouldn't have to encounter Dewey again
for twenty-four hours. Airmen returning home from
dangerous bombing runs will recognize the feeling.

It was in such a state of exultation one crisp and
twinkly March morning that I was delivering a paper to a
house half a block further on when Dewey – suddenly

twice his normal size and with truly unwarranted ferocity – came for me at speed from round the side of the McManuses' house. I remember thinking, in the microsecond for reflection that was available to me, that this was very unfair. It wasn't supposed to happen like this. This was my time of bliss.

Before I could meaningfully react, Dewey bit me hard on the leg just below the left buttock, knocking me to the ground. He then dragged me around for a bit – I remember my fingers scraping through grass – and then abruptly he released me and gave a confused, playful, woofy bark and bounded back into the border shrubbery whence he had come. Irate and comprehensively dishevelled, I waddled to the road to the nearest street light and took down my pants to see the damage. My jeans were torn, and on the fleshy part of my thigh there was a small puncture and a very little blood. It didn't actually hurt very much, but it came up the next day in a wonderful purply bruise, which I showed off in the boys' bathroom at school to many appreciative viewers, including Mr Groober, the strange, mute school janitor who was almost certainly an escapee from *some* place with high walls and who had never appeared quite this ecstatic about anything before, and I had to go to the doctor after school and get a tetanus shot, which I didn't appreciate a whole lot, as you can imagine.

Despite the evidence of my wound, the Haldemans refused to believe that their dog had gone for me. '*Dewey?*'

they laughed. 'Dewey wouldn't harm *any*one, honey. He wouldn't leave the property after dark. Why, he's afraid of his own shadow.' And then they laughed again. The dog that attacked me, they assured me, was some *other* dog.

Just over a week later, Dewey attacked Mrs Haldeman's mother, who was visiting from California. It had her down on the ground and was about to strip her face from her skull, which would have helped my case no end frankly. Fortunately for her, Mrs Haldeman came out just in time to save her mother and realize the shocking truth about her beloved pet. Dewey was taken away in a van and never seen again. I don't think anything has ever given me more satisfaction. I never did get an apology. However, I used to stick a secret booger in their paper every day.

At least rich people didn't move without telling you. My friend Doug Willoughby had a newspaper route at the more déclassé end of Grand Avenue, made up mostly of funny-smelling apartment buildings filled with dead-beats, shut-ins and people talking to each other through walls, not always pleasantly. All his buildings were gloomy and uncarpeted and all his corridors were so long and underlit that you couldn't see to the end of them, and so didn't know what was down there. It took resolution and nerve just to go in them. Routinely Willoughby would dis-cover that a customer had moved away (or been led off in handcuffs) without paying him, and Willoughby would have to make up the difference, for that's the way it worked. The *Register* never ended up out of pocket; only the paperboy did. Willoughby told me once that in his best week as a newspaper boy he made $4, and that included Christmas tips.

I, on the other hand, was steadily prospering, particularly when my bonus fines were factored in. Shortly before my twelfth birthday I was able to pay $102.12 in cash – a literally enormous sum; it took whole minutes to count it out at the cash register as it was mostly in small change – for a portable black and white RCA television with foldaway antenna. It was a new slimline model in whitish grey plastic, with the control knobs on top – an exciting innovation – and so extremely stylish. I carried it up to my room, plugged it in, switched it on and was seldom seen again around the house.

I took my dinner on a tray in my room each evening and scarcely ever saw my parents after that except on special occasions like birthdays and Thanksgiving. We bumped into each other in the hallway from time to time, of course, and occasionally on hot summer evenings I joined them on the screened porch for a glass of iced tea, but mostly we went our separate ways. So from that point our house was much more like a boarding house – a nice boarding house where the people got along well but respected and valued each other's privacy – than a family home.

All this seemed perfectly normal to me. We were never a terribly close family when I think back on it. At least we weren't terribly close in the conventional sense. My parents were always friendly, even affectionate, but in a slightly vague and distracted way. My mother was forever busy attacking collar stains or scraping potatoes off the oven walls – she was always attacking something – and my father was either away covering a sporting event for the paper or in his room reading. Very occasionally they went to a movie at the Varsity Theater – it showed Peter Sellers comedies from time to time, on which they quietly doted – or to the library, but mostly they stayed at home happily occupying different rooms.

Every night about eleven o'clock or a little after I would hear my father going downstairs to the kitchen to make a snack. My father's snacks were legendary. They took at least thirty minutes to prepare and required the most particular and methodical laying out of components – Ritz crackers, a large jar of mustard, wheat germ, radishes, ten Hydrox cookies, an enormous bowl of chocolate ice cream, several slices of luncheon meat, freshly washed lettuce, Cheez Whiz, peanut butter, peanut brittle, a hard-boiled egg or two, a small bowl of nuts, watermelon in season, possibly a banana – all neatly peeled, trimmed, sliced, cubed, stacked or layered as appropriate, and attractively arrayed on a large brown tray and taken away to be consumed over a period of hours. None of these snacks could have contained less than twelve thousand calories, at least 80 per cent of it in the form of cholesterol and saturated fats, and yet my father never gained an ounce of weight.

There was one other notable thing about my father's making of snacks. He was bare-assed when he made them. It wasn't, let me quickly add, that he thought being bare-assed somehow made for a better snack; it was just that he was bare-assed already. One of his small quirks was sleeping naked from the waist down. He believed that it was more comfortable, and healthful, to leave the bottom half of the body unencumbered at night, and so when in bed wore only a sleeveless T-shirt. And when he went downstairs late at night to concoct a snack he always went so attired (or unattired). Goodness knows what Mr and Mrs Bukowski next door must have thought as they drew their curtains and saw across the way (as surely they must) my father, bare-assed, padding about his kitchen, reaching into high cupboards and assembling the raw materials for his nightly feast.

Whatever dismay it may have caused next door, none of this was of any consequence in our house as everyone was in bed fast asleep (or in my case lying in the dark watching TV very quietly). But it happened that one night in about 1963, my father descended on a Friday night when my sister, unbeknown to him, was entertaining. Specifically, she and her good friends Nancy Ricotta and Wendy Spurgin were encamped in the living room with their boyfriends, watching television in the dark and swabbing each other's airways with their tongues (or so I have always imagined), when they were startled by a light coming on in the hallway above and the sound of my father descending the stairs.

As in most American homes, the living room in our house communicated with the rooms beyond by way of a doorless opening, in this case an arch about six feet wide, which meant that it offered virtually no privacy, so the sound of an approaching adult footfall was taken seriously. Instantly assuming positions of propriety, the six young people looked towards the entranceway just in time to see my father's lightly wobbling cheeks, faintly illumined by the ghostly flicker of television, passing through the hallway and proceeding onwards to the kitchen.

For twenty-five minutes they sat in silence, too mortified to speak, knowing that my father must return by the same route and that this time the encounter would be frontal.

Fortunately (insofar as such a word can apply here) my father must have peripherally noted them as he passed or heard voices or gasps or something, for when he returned with his tray he was snugly attired in my mother's beige raincoat, creating the impression that he was not only oddly depraved but a nocturnal cross-dresser as well. As he passed he mouthed a shy but pleasant good evening to the assembled party and disappeared back up the stairs.

It was about six months, I believe, before my sister spoke to him again.

* * *

Interestingly, at just about the time I acquired my television I realized that I didn't really like TV very much – or, to put it more accurately, didn't much like what was *on* TV, though I did like having the TV on. I liked the chatter and mindless laugh tracks. So mostly I left it babbling in the corner like a demented relative and read. I was at an age now where I read a lot, all the time. Once or twice a week I would descend to the living room, where there were two enormous (or so it seemed to me) built-in bookcases flanking the back window. These were filled with my parents' books, mostly hardback, mostly from the Book-of-the-Month Club, mostly from the 1930s and 1940s, and I would select three or four and take them up to my room.

I was happily indiscriminate in my selections because I had little idea which of the books were critically esteemed and which were popular tosh. I read, among much else, *Trader Horn*, *The Bridge of San Luis Rey*, *Our Hearts Were Young and Gay*, *Manhattan Transfer*, *You Know Me, Al*, *The Constant Nymph*, *Lost Horizon*, the short stories of Saki, several joky anthologies from Bennett Cerf, a thrilling account of life on Devil's Island called *Dry Guillotine*, and more or less the complete oeuvres of P. G. Wodehouse, S. S. Van Dine and Philo Vance. I had a particular soft spot for – and I believe may have been the last human being to read – *The Green Hat* by Michael Arlen with its wonderfully peerless names: Lady Pynte, Venice Pollen, Hugh Cypress, Colonel Victor Duck and the unsurpassable Trehawke Tush.

On one of these collecting trips, I came across, on a lower shelf, a Drake University Yearbook for 1936. Flipping through it, I discovered to my astonishment – complete and utter – that my mother had been homecoming queen that year. There was a picture of her on a float, radiant, beaming, slender, youthful, wearing a glittery tiara. I went with the book to the kitchen, where I found my father making coffee. 'Did you know Mom was homecoming queen at Drake?' I said.

'Of course.'

'How did *that* happen?'

'She was elected by her peers, of course. Your mom was quite a looker, you know.'

'Really?' It had never occurred to me that my mother looked anything except motherly.

'Still is, of course,' he added chivalrously.

I found it astounding, perhaps even a little out of order, that other people might find my mother attractive or desirable. Then I quite warmed to the idea. My mother had been a beauty. Imagine.

I put the book back. On the same section of shelf were eight or nine books entitled *Best Sports Stories of 1950* and so on for nearly every year up to the present, each consisting of thirty or forty of the best sports articles of that year as chosen by somebody well known like Red Barber. Each of these volumes contained a piece of work – in some cases two pieces – by my dad. Often he was the only provincial journalist included. I sat down on the window

seat between the bookcases and read several of them right
there. They were wonderful. They really were. It was just one
bright line after another. One I recall recorded how
University of Iowa football coach Jerry Burns ranged up and
down the sidelines in dismay as his defensive team haplessly
allowed Ohio State to score touchdowns at will. 'It was a
case of the defence fiddling while Burns roamed,' he wrote,
and I was amazed to realize that the bare-assed old fool was
capable of such flights of verbal scintillation.

In light of these heartening discoveries, I amended
the Thunderbolt Kid story at once. I *was* their biological
offspring after all – and pleased to be so. Their genetic
material was my genetic material and no mistake. It
turned out, on further consideration, that it must have
been my father, not I, who had been dispatched to Earth
from Planet Electro to preserve and propagate the interests
of King Volton and his doomed race. That made vastly
more sense when I thought about it. What better-sound-
ing place, after all, for a superhero to grow up in than
Winfield, Iowa? *That*, surely, was where the Thunderbolt
Kid was intended to come from.

Unfortunately, I realized now, my father's space
capsule had suffered a hard landing, and my father had
received a concussive bump, which had wiped his
memory clean and left him with one or two slightly
strange habits – a crippling cheapness and a dis-
inclination to wear underpants after dark being the
principal ones – and spent his whole life tragically

241

unaware that he had the innate capacity to summon up superpowers. Instead, it was left to his youngest son to make that discovery. That was why I needed special clothes to assume my Electron powers. I was an Earthling by birth, so I didn't come by these super-gifts naturally. I required the Sacred Jersey of Zap for that.

Of course. It all made sense now. This story just got better and better, in my view.

Chapter 10
DOWN ON THE FARM

MASON CITY, IOWA – A pretty blonde bride's playful tickling of her husband to get him out of bed to milk the cows led swiftly to tragedy early Tuesday. Mrs. Jennie Becker Brunner, 22, said through her tears in Cerro Gordo County jail cell here late in the day that she shot and killed her husband, Sam Brunner, 26, with his .45 caliber U.S. Army Colt pistol. Mrs. Brunner said she and her husband quarreled after she tickled him under the arm to get him out of bed.

– *Des Moines Register*, 19 November 1953

GIVE OR TAKE the occasional ticklish murder, Iowa has always been a peaceful and refreshingly unassertive place. In the one hundred and sixty years or so that it has been a state, only one shot has been officially fired in anger on Iowa soil, and even that wasn't very angry. During the Civil War, a group of Union soldiers, for reasons that I believe are now pretty well forgotten, discharged a cannonball across the state line into Missouri. It landed in a field on the other side and dribbled harmlessly to a halt. I shouldn't be surprised if the Missourians put it on a wagon and brought it back. In any case, nobody was hurt. This was not simply the high point in Iowa's military history, it was the only point in it.

Iowa has always been proudly middling in all its affairs. It stands in the middle of the continent, between the two mighty central rivers, the Missouri and Mississippi, and throughout my childhood always ranked bang in the middle of everything – size, population, voting preferences, order of entry into the Union. We were slightly wealthier, a whole lot more law-abiding, and more literate and better educated than the national average, and ate more Jell-O (a lot more – in fact, to be

245

completely honest, we ate all of it), but otherwise have never been too showy at all. While other states of the Midwest churned out a more or less continuous stream of world-class worthies – Mark Twain, Abraham Lincoln, Ernest Hemingway, Thomas Edison, Henry Ford, F. Scott Fitzgerald, Charles Lindbergh – Iowa gave the world Donna Reed, Wyatt Earp, Herbert Hoover and the guy who played Fred Mertz on *I Love Lucy*.

Iowa's main preoccupations have always been farming and being friendly, both of which we do better than almost anyone else, if I say so myself. It is the quintessential farm state. Everything about it is perfect for growing things. It occupies just 1.6 per cent of the country's land area, but contains 25 per cent of its Grade A topsoil. That topsoil is three feet deep in most places, which is apparently pretty deep. Stride across an Iowa farm field and you feel as if you could sink in up to your waist. You will certainly sink in up to your ankles. It is like walking around on a very large pan of brownies. The climate is ideal, too, if you don't mind shovelling tons of snow in the winter and dodging tornadoes all summer. By the standards of the rest of the world, droughts are essentially unknown and rainfall is distributed with an almost uncanny beneficence – heavy enough to give a healthful soaking when needed but not so much as to pummel seedlings or wash away nutrients. Summers are long and agreeably sunny, but seldom scorching. Plants love to grow in Iowa.

It is in consequence one of the most maximally farmed landscapes on earth. Someone once calculated that if Iowa contained nothing but farms, each of one hundred and sixty acres (presumably the optimal size for a farm), there would be room for 225,000 of them. In 1930, the peak year for farm numbers, there were 215,361 farms in the state – not far off the absolute maximum. The number is very much smaller these days because of the relentless push of amalgamation, but 95 per cent of Iowa's landscape is still farmed. The remaining small fraction is taken up by highways, woods, a scattering of lakes and rivers, loads of little towns and a few smallish cities, and about twelve million Wal-Mart parking lots.

I remember reading once at the State Fair that Iowa's farms produced more in value each year than all the diamond mines in the world put together – a fact that fills me with pride still. It remains number one in the nation for the production of corn, eggs, hogs and soybeans, and is second in the nation in total agricultural wealth, exceeded only by California, which is three times the size. Iowa produces one tenth of all America's food and one tenth of all the world's maize. Hooray.

And when I was growing up all this was as good as it has ever been. The 1950s has often been called the last golden age of the family farm in America, and no place was more golden than Iowa, and no spot had a lovelier glint than Winfield, the trim and cheerful little town in the southeast corner of the state, not far from the

Mississippi River, where my father had grown up and my grandparents lived.

I loved everything about Winfield – its handsome Main Street, its imperturbable tranquillity, its lapping cornfields, the healthful smell of farming all around. Even the name was solid and right. Lots of towns in Iowa have names that sound slightly remote and lonesome and perhaps just a little inbred – Mingo, Pisgah, Tingley, Diagonal, Elwood, Coon Rapids, Ricketts – but in this green and golden corner of the state the town names were dependably worthy and good: Winfield, Mount Union, Columbus Junction, Olds, Mount Pleasant, the un-beatably radiant Morning Sun.

My grandfather was a rural route mailman by trade, but he owned a small farm on the edge of town. He rented out the land to other farmers, except for three or four acres that he kept for orchards and vegetables. The property included a big red barn and what seemed to me like huge lawns on all sides. The back of the house was dominated by an immense oak tree with a white bench encircling it. It seemed always to have a private breeze running through its upper branches. It was the coolest spot in a hundred miles. This was where you sat to shuck peas or trim green beans or turn a handle to make ice cream at the tranquil, suppertime end of the day.

My grandparents' house was very neat and small – it had just two bedrooms, one upstairs and one down – but was exceedingly comfortable and always seemed spacious to me. Years later I went back to Winfield and was astounded at how tiny it actually was.

From a safe distance, the barn looked like the most fun place in the world to play. It hadn't been used for years except to store old furniture and odds and ends that would never be used again. It was full of doors you could swing on and secret storerooms and ladders leading up to dark haymows. But it was actually awful because it was filthy and dark and lethal and every inch of it smelled. You couldn't spend five minutes in my grandfather's barn without banging your shins on some piece of unyielding machinery, cutting your arm on an old blade, coming into contact with at least three different types of ancient animal shit (all years old but still soft in the middle), banging your head on a nail-studded beam and recoiling into a mass of sticky cobwebs, getting snagged from the nape of your neck to the top of your buttocks on a strand of barbed wire, quilling yourself all over with splinters the size of toothpicks. The barn was like a whole-body workout for your immune system.

The worst fear of all was that one of the heavy doors would swing shut behind you and you would be trapped for ever in a foul smelly darkness, too far from the house for your plaintive cries to be heard. I used to imagine my family sitting round the dinner table saying, 'Well, I wonder whatever became of old Billy. How long has it been now? Five weeks? Six? He'd sure love this pie, wouldn't he? I'll certainly have another piece if I may.'

Even scarier were the fields of corn that pressed in on all sides. Corn doesn't grow as tall as it used to because it's been hybridized into a more compact perfection, but it shot up like bamboo when I was young, reaching heights of eight feet or more and filling 56,290 square miles of Iowa countryside with a spooky, threatening rustle by the dryish late end of summer. There is no more anonymous, mazelike, unsettling environment, especially to a dim, smallish human, than a field of infinitely identical rows of tall corn, each – including the diagonals – presenting a prospect of endless vegetative hostility. Just standing on the edge and peering in, you knew that if you ventured more than a few feet into a cornfield you would never come out. If a ball you were playing with dropped into a cornfield, you just left it, wrote it off, and went inside to watch TV.

So I didn't play alone much at Winfield. Instead I spent a lot of time following my grandfather around. He seemed to like the company. We got along very well. My grandfather was a quiet man, but always happy to explain what he was doing and glad to have someone who could pass him an oil can or a screwdriver. His name was Pitt Foss Bryson, which I thought was the best name ever. He was the nicest man in the world after Ernie Banks.

He was always rebuilding something – a lawnmower or washing machine; something with fan belts and blades and lots of swiftly whirring parts – and always cutting himself fairly spectacularly. At some point, he would fire the thing up, reach in to make an adjustment and almost immediately go, 'Dang!' and pull out a bloody, slightly shredded hand. He would hold it up before him for some time, wiggling the fingers, as if he didn't quite recognize it.

'I can't see without my glasses,' he would say to me at length. 'How many fingers have I got here?'

'Five, Grandpa.'

'Well, *that's* good,' he'd say. 'Thought I might have lost one.' Then he'd go off to find a bandage or piece of rag.

At some point in the afternoon, my grandmother would put her head out the back door and say, 'Dad, I need you to go uptown and get me some rutabaga.' She always called him Dad, even though he had a wonderful name and he wasn't her father. I could never understand that. She always needed him to get rutabaga. I never understood that either since I don't remember any of us ever being served it. Maybe it was a code word for prophylactics or something.

Going uptown was a treat. It was only a quarter of a mile or so, but we always drove, sitting on the high bench seat of my grandfather's Chevy, which made you feel slightly regal. Uptown in Winfield meant Main Street, a two-block stretch of retail tranquillity sporting a post office, two banks, a couple of filling stations, a tavern, a newspaper office, two small grocer's, a pool hall and a variety store.

The last stop on every shopping trip was a corner grocer's called Benteco's, where they had a screen door that *kerboinged* and *bammed* in a deeply satisfying manner, and made every entrance a kind of occasion. At Benteco's I was always allowed to select two bottles of NeHi brand pop – one for dinner, one for afterwards when we were playing cards or watching Bilko* or Jack Benny on TV. NeHi was the pop of small towns – I don't know why – and it had the intensest flavour and most vivid colours of any products yet cleared by the Food and Drug Administration for human consumption. It came in six select flavours – grape, strawberry, orange, cherry, lime-lemon (never 'lemon-lime') and root beer – but each was so potently flavourful that it made your eyes water like an untended sprinkler, and so sharply carbonated that it was like swallowing a thousand tiny razor blades. It was wonderful.

The NeHi at Benteco's was kept in a large, blue, very chilly cooler, like a chest freezer, in which the bottles hung by their necks in rows. To get to a particular bottle usually required a great deal of complicated manoeuvring, transferring bottles out of one row and into another in order to get the last bottle of grape, say. (Grape was the one flavour that could actually make you hallucinate; I once saw to the edge of the universe while drinking grape NeHi.) The process was great fun if it was you that was doing the selecting (especially on a hot day when you could bask in

*I know it was never actually called *Bilko*. It was *You'll Never Get Rich* and then later *The Phil Silvers Show*. But we called it *Bilko*. Everybody did. It was only on for four years.

the cooler's moist chilled air) and a torment if you had to wait on some other kid.

The other thing I did a lot in Winfield was watch TV. My grandparents had the best chair for watching television – a beige leatherette recliner that was part fairground ride, part captain's seat from a spaceship, and all comfort. It was a thing of supreme beauty and utility. When you pulled the lever you were thrust – flung – into a deep recline mode. It was nearly impossible to get up again, but it didn't matter because you were so sublimely comfortable that you didn't want to move. You just lay there and watched the TV through splayed feet.

My grandparents could get seven stations on their set – we could only get three in Des Moines – but only by turning the roof aerial, which was manipulated by means of a crank on the outside back wall of the house. So if you wanted to watch, say, KTVO from Ottumwa, my grandfather had to go out and turn the crank slightly one way, and if you wanted WOC from the Quad Cities he turned it another, and KWWI in Waterloo another way still, in each case responding to instructions shouted through a window. If it was windy or there was a lot of solar activity, he sometimes had to go out eight or nine times during a programme. If it was one of my grandmother's treasured shows, like As the World Turns or Queen for a Day, he generally just stayed out there in case an aeroplane flew over and made everything lapse into distressing waviness at a critical moment. He was the most patient man that ever lived.

I watched a lot of television in those days. We all did. By 1955, the average American child had watched five thousand hours of television, up from zero hours five years earlier. My favourite programmes were, in no particular order, *Zorro*, *Bilko*, *Jack Benny*, *Dobie Gillis*, *Love That Bob*, *Sea Hunt*, *I Led Three Lives*, *Circus Boy*, *Sugarfoot*, *M Squad*, *Dragnet*, *Father Knows Best*, *The Millionaire*, *Gunsmoke*, *Robin Hood*, *The Untouchables*, *What's My Line?*, *I've Got a Secret*, *Route 66*, *Topper* and *77 Sunset Strip*, but really I would watch anything.

My favourite of all was the *Burns and Allen Show* starring George Burns and Gracie Allen. I was completely enchanted with it because I loved the characters and their names – Blanche Morton, Harry Von Zell – and because George Burns and Gracie Allen were, in my view, the funniest double act ever. George had a deadpan manner and Gracie always got the wrong end of every stick. George had a television in his den on which he could watch what his neighbours were up to without their knowing it, which I thought was just a brilliant notion and one that fed many a private fantasy, and he often stepped out of the production to talk directly to the audience about what was going on. The whole thing was years ahead of its time. I've never met another human being who even remembers it, much less doted on it.

Nearly every summer evening just before six o'clock we would walk uptown (all movement towards the centre was known as going uptown) to some shady church lawn and take part in a vast potluck supper, presided over by armies of immense, chuckling women who had arms and necks that sagged in an impossible manner, like really wet clothes. They were all named Mabel and they all suffered greatly from the heat, though they never complained and never stopped chuckling and being happy. They spent their lives shooing flies from food with spatulas (setting their old arms a-wobbling in a hypnotizing manner), blowing wisps of stray hair out of their faces, and making sure that no human being within fifty yards failed to have a heaped paper plate of hearty but deeply odd food – and dinners in the 1950s, let me say, were odd indeed. The main courses at these potluck events nearly always consisted of a range of meatloafs, each about the size of a V8 engine, all of them glazed and studded with a breathtaking array of improbable ingredients from which they drew their names – Peanut Brittle 'n' Cheez Whiz Upside Down Spam Loaf and that sort of thing. Nearly all of them had at least one ''n'' and an 'upside down' in their names somewhere. There would be perhaps twenty of these. The driving notion seemed to be that no dish could be too sweet or too strange and that all foods automatically became superior when upended.

'Hey, Dwayne, come over here and try some of this Spiced Liver 'n' Candy Corn Upside Down Casserole,' one of the Mabels would say. 'Mabel made it. It's delicious.'

'Upside *down*?' Dwayne would remark with a dry

look that indicated a quip was coming. 'What happened – she drop it?'

'Well, I don't know. Maybe she did,' Mabel would reply, chuckling. 'You want chocolate gravy with that or biscuit gravy or peanut butter 'n' niblets gravy?'

'Hey, how about a little of all three?'

'You got it!'

The main dishes were complemented by a table of brightly coloured Jell-Os, the state fruit, each containing further imaginative components – marshmallows, pretzels, fruit chunks, Rice Krispies, Fritos corn chips, whatever would maintain its integrity in suspension – and you had to take some of each of these, too, though of course you wanted to because it all looked so tasty. Then came at least two big tables carrying tubs and platters of buttery mashed potatoes, bacon swimming in baked beans, creamed vegetables, devilled eggs, cornbreads, muffins, heavy-duty biscuits and a dozen types of cole slaw. By the time all these were loaded on to your paper plate, it weighed twelve pounds and looked, as my father once described it, distinctly post-operative. But there was no resisting the insistent blandishments of the many Mabels.

Everyone for miles came to the suppers. It didn't matter what the denomination of the church was. Everybody came. Everyone in town was practically Methodist anyway, even the Catholics. (My grandparents, for the record, were Lutherans.) It wasn't about religion; it was about sociable eating in bulk.

'Now don't forget to leave room for dessert,' one of the Mabels would say as you staggered off with your plate, but you didn't have to be reminded of that for the desserts were fabulous and celebrated, the best part of all. They were essentially the same dishes, but with the meat removed.

On the few nights when we weren't at a church social, we had enormous meals at my grandparents' house, often on a table carried out to the lawn. (It seemed important to people in those days to share dinner with as many insects as possible.) Uncle Dee would be there, of course, burping away, and Uncle Jack from Wapello, who was notable for never managing to finish a sentence.

'I tell you what they ought to do,' he would say in the midst of a lively discussion, and someone else more assertive would interject a comment and nobody would ever hear what Jack thought. 'Well, if you ask me,' he'd say, but nobody ever did. Mostly they sat around talking about surgical removals and medical conditions – goitres and gallstones, lumbago, sciatica, water on the knee – that don't seem to exist much any more. They always seemed so old to me, and slow, so glad to sit down.

But they sure were good natured. If we had a guest from beyond the usual family circle somebody would always bring out the dribble glass and offer the guest a drink. The dribble glass was the funniest thing I had ever seen. It was a fancy-looking, many-faceted drinking glass – exactly the sort of glass that you would give to an

257

honoured guest – that appeared to be perfectly normal, and indeed was perfectly normal so long as you didn't tilt it. But cut into the facets were tiny, undetectable slits, ingeniously angled so that each time the glass was inclined to the mouth a good portion of the contents dribbled out in a steady run on to the victim's chest.

There was something indescribably joyous about watching an innocent, unaware person repeatedly staining him- or herself with cranberry juice or cherry Kool-Aid (it was always something vividly coloured) while twelve people looked on with soberly composed expressions. Eventually, feeling the seepage, the victim would look down and cry, 'Oh, my golly!' and everyone would burst out laughing.

I never knew a single victim to get angry or dismayed when they discovered the prank. Their best white shirt would be ruined, they would look as if they had been knifed in the chest, and they would laugh till their eyes streamed. God, but Iowans were happy souls.

Winfield always had more interesting weather than elsewhere. It was always hotter, colder, windier, noisier, sultrier, more punishing and emphatic than weather elsewhere. Even when the weather wasn't actually doing anything, when it was just muggy and limp and still on an August afternoon, it was more muggy and limp than anywhere else you have ever been, and so still that you could hear a clock ticking in a house across the street.

Because Iowa is flat and my grandparents lived on the very edge of town, you could see everything meteorological long before it got there. Storms of towering majesty often lit the western skies for two or three hours before the first drops of rain fell in Winfield. They talk about big skies in the western US, and they may indeed have them, but you have never seen such lofty clouds, such towering anvils, as in Iowa in July.

The greatest fury in Iowa – in the Midwest – is tornadoes. Tornadoes are not often seen because they tend to be fleeting and localized and often they come at night, so you lie in bed listening to a wild frenzy outside knowing that a tornado's tail could dip down at any moment and blow you and your cosy tranquillity to pieces. Once my grandparents were in bed when they heard a great roaring, like a billion hornets as my grandfather described it, going right past their house. My grandfather got up and peered out the bedroom window but couldn't see a thing and went back to bed. Almost at once the noise receded.

In the morning, he stepped outside to fetch in the newspaper and was surprised to find his car standing in the open air. He was sure he *had* put it away as usual the night before. Then he realized he had put the car away, but the garage was gone. The car was standing on its concrete floor. It didn't have a scratch on it. Nothing of the garage was ever seen again. Looking closer, he discovered a track of destruction running along one side of the house. A bed of shrubs that had stood against the house, in front of the bedroom window, had been obliterated utterly, and he realized that the blackness he had peered into the night before was a wall of tornado passing on the other side of the glass just an inch or two beyond his nose.

Just once I saw a tornado when I was growing up. It was moving across the distant horizon from right to left, like a killer apostrophe. It was about ten miles off and therefore comparatively safe. Even so it was unimaginably powerful. The sky everywhere was wildly, unnaturally dark and heavy and low, and every wisp of cloud in it, from every point in the compass, was being sucked into the central vortex as if being pulled into a black hole. It was like being present for the end of the world. The wind, steady and intense, felt oddly as if it was not pushing from behind, but pulling from the front, like the insistent draw of a magnet. You had to fight not to be pulled forward. All that energy was being focused on a single finger of whirring destruction. We didn't know it at the time, but it was killing people as it went.

For a minute or two the tornado paused in its progress and seemed to stand on one spot.

'That could mean it's coming towards us,' my father remarked to my grandfather.

I took this to mean that we would all now get in our cars and drive like hell in a contrary direction. That was the option I planned to vote for if anyone asked for a show of hands.

But my grandfather merely said, 'Yup. Could be,' and looked completely undisturbed.

'Ever seen a tornado up close, Billy?' my father said to me, smiling weirdly.

I stared at him in amazement. Of course not and I didn't want to. This business of not ever being frightened of anything was easily the most frightening thing about adults in the Fifties.

'What do we do if it's coming this way?' I asked in a pained manner, knowing I was not going to enjoy the answer.

'Well, that's a good question, Billy, because it's very easy to flee from one tornado and drive straight into another. Do you know, more people die trying to get out of the way of tornadoes than from any other cause?' He turned to my grandfather. 'Do you remember Bud and Mabel Weidermeyer?'

My grandfather nodded with a touch of vigour, as if to say, *Who could forget it?* 'They should have known better than to try to outrun a tornado on foot,' my grandfather said. 'Especially with Bud's wooden leg.'

'Did they ever find that leg?'

'Nope. Never found Mabel either. You know, I think it's moving again.'

He indicated the tornado and we all watched closely. After a few moments it became apparent that it had indeed resumed its stately march to the east. It wasn't coming towards us after all. Very soon after that, it lifted from the ground and returned into the black clouds above it, as if being withdrawn. Almost at once the wind dropped. My father and grandfather went back in the house looking slightly disappointed.

The next day we drove over and had a look at where it had gone and there was devastation everywhere – trees and power lines down, barns blown to splinters, houses half vanished. Six people died in the neighbouring county. I expect none of them were worried about the tornado either.

What I particularly remember of Winfield is the coldness of the winters. My grandparents were very frugal with the heat in their house and tended to turn it all but off at night, so that the house never warmed up except in the kitchen when a big meal was being cooked, like at Thanksgiving or Christmas, when it took on a wonderful steamy warmth. But otherwise it was like living in an Arctic hut. The upstairs of their house was a single long room, which could be divided into two by a pull-across curtain. It had no heating at all and the coldest linoleum floor in history.

But there was one place even colder: the sleeping porch. The sleeping porch was a slightly rickety, loosely enclosed porch on the back of the house that was only notionally separate from the outside world. It contained an ancient sagging bed that my grandfather slept in in the summer when the weather was unbearably hot. But sometimes in the winter when the house was full of guests it was pressed into service, too.

The only heat the sleeping porch contained was that of any human being who happened to be out there. It couldn't have been more than one or two degrees warmer than the world outside – and outside was perishing. So to sleep on the sleeping porch required preparation. First, you put on long underwear, pyjamas, jeans, a sweatshirt, your grandfather's old cardigan and bathrobe, two pairs of woollen socks on your feet and another on your hands, and a hat with ear flaps tied beneath the chin. Then you climbed into bed and were immediately covered with a dozen bed blankets, three horse blankets, all the household overcoats, a canvas tarpaulin and a piece of old carpet. I'm not sure that they didn't lay an old wardrobe on top of that, just to hold everything down. It was like sleeping under a dead horse. For the first minute or so it was unimaginably cold, shockingly cold, but gradually your body heat seeped in and you became warm and happy in a way you would not have believed possible only a minute or two before. It was bliss.

Or at least it was until you moved a muscle. The warmth, you discovered, extended only to the edge of your skin and not a micron further. There wasn't any possibility of shifting positions. If you so much as flexed a finger or bent a knee, it was like plunging them into liquid nitrogen. You had no choice but to stay totally immobilized. It was a strange and oddly wonderful experience – to be poised so delicately between rapture and torment.

It was the serenest, most peaceful place on earth. The view from the sleeping porch through the big broad window at the foot of the bed was across empty dark fields to a town called Swedesburg, named for the nationality of its founders, and known more informally as Snooseville from the pinches of tobacco that the locals used to pack into their mouths as they went about their business. Snoose was a homemade mixture of tobacco and salt which was kept embedded between cheek and gum where the nicotine could be slowly and steadily absorbed. It was topped up hourly and kept in permanently. Some people, my father told me, even put in a fresh wad at bedtime.

I had never been to Swedesburg. There was no reason to go – it was just a small collection of houses – but at night in winter with its distant lights it was like a ship far out at sea. I found it peaceful and somehow comforting to see their lights, to think that all the citizens of Snooseville were snug in their homes and perhaps looking over at us in Winfield and deriving comfort in turn. My father told me that when he was a boy the people of Snooseville still spoke Swedish at home. Some of them could barely speak English at all. I loved that, too – the idea that it was a little outpost of Sweden over there, that they were all sitting around eating herring and black bread and saying, 'Oh, ja!' and just being happily Swedish in the middle of the

American continent. When my dad was young if you drove across Iowa you would regularly come across towns or villages where all the inhabitants spoke German or Dutch or Czech or Danish or almost any other tongue from northern and central Europe.

But those days had long since passed. In 1916, as the shadow of the Great War made English-speaking people suspicious of loyalties, a governor of Iowa named William L. Harding decreed that henceforth it would be a crime to speak any foreign language in schools, at church, or even over the telephone in the great state of Iowa. There were howls that people would have to give up church services in their own languages, but Harding was not to be moved. 'There is no use in anyone wasting his time praying in other languages than English,' he responded. 'God is listening only to the English tongue.'

One by one the little linguistic outposts faded away. By the 1950s they were pretty much gone altogether. No one would have guessed it at the time, but the small towns and family farms were soon to become likewise imperilled.

In 1950, America had nearly six million farms. In half a century almost two thirds of them vanished. More than half the American landscape was farmed when I was a boy; today, thanks to the spread of concrete, only 40 per cent is – a severe decline in a single lifetime.

I was born into a state that had two hundred thousand farms. Today the number is much less than half that and falling. Of the seven hundred and fifty thousand people who lived on farms in the state in my boyhood, half a million – two in every three – have gone. The process has been relentless. Iowa's farm population fell by 25 per cent in the 1970s and by 35 per cent more in the 1980s. Another hundred thousand people were skimmed away in the 1990s. And the people left behind are old. In 1988, Iowa had more people who were seventy-five or older than five or younger. Thirty-seven counties out of ninety-nine – getting on for half – recorded more deaths than births.

It's an inevitable consequence of greater efficiency and continuous amalgamation. Increasingly the old farms clump together into superfarms of three thousand acres or more. By the middle of this century, it is thought, the number of farms in Iowa could drop to as low as ten thousand. That's not much of a rural population in a space the size of England.

Without a critical mass of farmers, most small towns in Iowa have pretty well died. Drive anywhere in the state these days and what you see are empty towns, empty roads, collapsing barns, boarded farmhouses. Everywhere you go it looks as if you have just missed a terrible contagion, which in a sense I suppose you have. It's the same story in Illinois, Kansas and Missouri, and even worse in Nebraska and the Dakotas. Wherever there were once small towns, there are now empty main streets.

Winfield is barely alive. All the businesses on Main

Street – the dime store, the pool hall, the newspaper office, the banks, the grocery stores – long ago disappeared. There is nowhere to buy NeHi pop even if it still existed. You can't purchase a single item of food within the town limits. My grandparents' house is still there – at least it was the last time I passed – but its barn is gone and its porch swing and the shade tree out back and the orchard and everything else that made it what it was.

The best I can say is that I saw the last of something really special. It's something I seem to say a lot these days.

Chapter 11

WHAT, ME WORRY?

LIES IN MORGUE 17 HOURS – ALIVE

ATLANTA, GA. (UP) – An elderly woman taken to a funeral home for embalming opened her eyes 17 hours after arriving and announced: 'I'm not dead.'

W. L. Murdaugh of Murdaugh Brothers funeral home here said two of his employees were made almost speechless.

The woman, listed as Julia Stallings, 70, seemed dazed after her long coma ended Sunday night, but otherwise appeared in good condition, Murdaugh said.

– *Des Moines Tribune*, 11 May 1953

Collier's

15c

August 5, 1950

HIROSHIMA, U.S.A.

Can Anything Be Done About It?

THE ONLY TIME I HAVE EVER broken a bone was also the first time I noticed that adults are not entirely to be counted on. I was four years old and playing on Arthur Bergen's jungle gym when I fell off and broke my leg.

Arthur Bergen lived up the street, but was at the dentist or something when I called, so I decided to have a twirl on his new jungle gym before heading back home.

I don't remember anything at all about the fall, but I do remember very clearly lying on damp earth, the jungle gym now above and around me and seeming awfully large and menacing all of a sudden, and not being able to move my right leg. I remember also lifting my head and looking down my body to my leg which was bent at an unusual – indeed, an entirely novel – angle. I began to call steadily for help, in a variety of tones, but no one heard. Eventually I gave up and dozed a little.

At some point I opened my eyes and a man with a uniform and a peaked cap was looking down at me. The sun was directly behind him so I couldn't see his face; it was just a hatted darkness inside a halo of intense light.

'You all right, kid?' he said.

'I've hurt my leg.'

He considered this for a minute. 'You wanna get your mom to put some ice on it. Do you know some people named . . .' – he consulted a clipboard – '. . . Maholovich?'

'No.'

He glanced at the clipboard again. 'A. J. Maholovich. 3725 Elmwood Drive.'

'No.'

'Doesn't ring a bell at all?'

'No.'

'This is Elmwood Drive?'

'Yes.'

'OK, kid, thanks.'

'It really hurts,' I said. But he was gone.

I slept a little more. After a while Mrs Bergen pulled into their driveway and came up the back steps with bags of groceries.

'You'll catch a chill down there,' she said brightly as she skipped past.

'I've hurt my leg.'

She stopped and considered for a moment. 'Better get up and walk around on it. That's the best thing. Oh, there's the phone.' She hurried into the house.

I waited for her to come back but she didn't. 'Hello,' I croaked weakly now. 'Help.'

Bergen's little sister, who was small and therefore stupid and unreliable, came and had a critical look at me.

'Go and get your mom,' I said. 'I'm hurt.'

She looked at my leg with comprehension if not compassion. 'Owie,' she said.

'Yes, owie. It really hurts.'

She wandered off, saying, 'Owie, owie,' but evidently took my case no further.

Mrs Bergen came out after some time with a load of washing to hang.

'You must really like it down there,' she chuckled.

'Mrs Bergen, I think I've really hurt my leg.'

'On that little jungle gym?' she said, with good-natured scepticism, but came closer to look at me. 'I don't think so, honey.' And then abruptly: 'Christamighty! Your leg! It's backwards!'

'It hurts.'

'I bet it does, I bet it does. You wait right there.'

She went off.

Eventually, after quite some time, Mr Bergen and my parents pulled up in their respective cars at more or less the same moment. Mr Bergen was a lawyer. I could hear him talking to them about liability as they came up the steps. Mr Bergen was the first to reach me.

'Now you do understand, Billy, that technically you were trespassing . . .'

They took me to a young Cuban doctor on Woodland Avenue and *he* was in a panic. He started making exactly the kind of noises Desi Arnaz made in *I Love Lucy* when Lucy did something really bone-headed – only he was doing this over my leg. 'I don' thin' I can do this,' he said, and looked at them beseechingly. 'It's a really bad break. I mean look at it. Wow.'

I expect he was afraid he would be sent back to Cuba. Eventually he was prevailed upon to set the break. For the next six weeks my leg remained more or less backwards. The moment they cut off the cast, the leg spun back into position and everyone was pleasantly surprised. The doctor beamed. 'Tha's a bit of luck!' he said happily.

Then I stood up and fell over.

'Oh,' the doctor said and looked troubled again. 'Tha's not good, is it?'

He thought for a minute and told my parents to take me home and to keep me off the leg for the rest of the day and overnight and see how it was in the morning.

'Do you think it will be all right then?' asked my father.

'I've no idea,' said the doctor.

The next morning I got up and stepped gingerly on to my wounded leg. It felt OK. It felt good. I walked around. It was fine. I walked a little more. Yes, it was definitely fine.

I went downstairs to report this good news and found my mother bent over in the laundry room sorting through clothes.

'Hey, Mom, my leg's fine,' I announced. 'I can walk.'

'Oh, that's good, honey,' she said, head in the dryer. 'Now where's that other sock?'

* * *

It wasn't that my mother and father were indifferent to their children's physical well-being by any means. It was just that they seemed to believe that everything would be fine in the end and they were always right. No one ever got lastingly hurt in our family. No one died. Nothing ever went seriously wrong – and not much went wrong in our town or state either, come to that. Danger was something that happened far away in places like Matsu and Quemoy, and the Belgian Congo, places so distant that nobody was really quite sure where they were.

It's hard for people now to remember just how enormous the world was back then for everybody, and how far away even fairly nearby places were. When we called my grandparents long distance on the telephone in Winfield, something we hardly ever did, it sounded as if they were speaking to us from a distant star. We had to shout to be heard and plug a finger in an ear to catch their faint, tinny voices in return. They were only about a hundred miles away, but that was a pretty considerable distance even well into the 1950s. Anything further – beyond Chicago or Kansas City, say – quickly became almost foreign. It wasn't just that Iowa was far from everywhere. Everywhere was far from everywhere.

America was especially blessed in this regard. We had big buffering oceans to left and right and no neighbours to worry us above or below, so there wasn't any need to be fearful about anything ever. Even world wars barely affected our home lives. During the Second World War, when the film mogul Jack Warner realized that from the air his Hollywood studio was indistinguishable from a nearby aircraft factory, he had a giant arrow painted on the roof above the legend 'Lockheed That-Away!' to steer Japanese bombers safely away from some of the valuable stars who didn't go to war (and that included Gary Cooper, Bob Hope, Fred MacMurray, Frank Sinatra, John Garfield, Gene Kelly, Alan Ladd, Danny Kaye, Cary Grant, Bing Crosby, Van Johnson, Dana Andrews, Ronald Reagan and John Wayne, among many other valiant heroes who helped America to act its way to victory) and towards the correct target.

No one ever knew whether Warner was in earnest with his sign or not, but it didn't really matter because no one seriously expected (at least not after the first jittery days of the war) that the Japanese would attack the US mainland. At the same time, on the other side of the country, when a Congressman grew concerned for the welfare of rooftop sentries at the Capitol Building who didn't ever seem to stir from their positions or enjoy a moment's relief, he was quietly informed that they were in fact dummies and that their anti-aircraft guns were wooden models. There was no point in wasting men and munitions on a target that was never going to be hit, even if it was the headquarters of the United States government.

For the record there was one manned attack on the American mainland. In 1942, a pilot named Nobuo Fujita took to the air from coastal waters off Oregon in a specially modified seaplane that was brought there aboard a submarine. Fujita's devious goal was to drop incendiary bombs on Oregon's forests, starting large-scale fires that would, if all went to plan, rage out of control and engulf much of the West Coast, killing hundreds and leaving Americans weeping and demoralized at the thought of all that damage caused by one little squinty-eyed man in a plane. In the event, the bombs either puttered out or caused only localized fires of no consequence.

The Japanese also, over a period of months, launched into the prevailing winds across the Pacific some nine thousand large paper balloons, each bearing a thirty-pound bomb timed to go off forty hours after launch – the length of time calculated that it would take to cross the Pacific to America. These managed to blow up a small number of curious souls whose last earthly utterance was something along the lines of 'Now what the heck do you suppose this is?', but otherwise did almost no damage, though one made it as far as Maryland.

Then in the Cold War years all this comfortable security abruptly vanished as the Soviet Union developed long-range ballistic missiles to match our own. Suddenly we were in a world where something horribly destructive could drop on us at any moment without warning wherever we were. This was a startling and unsettling notion, and we responded to it in a quintessentially 1950s way. We got excited about it.

For a number of years you could hardly open a

magazine without learning of some new destructive marvel that could wipe us all out in a twinkle. An artist named Chesley Bonestell specialized in producing sumptuously lifelike illustrations of manmade carnage, showing warhead-laden rockets streaking gorgeously (excitingly!) across American skies or taking off from giant space stations on a beautifully lit, wondrously imagined Moon en route to an explosive attack on planet Earth.

The thing about Bonestell's paintings was that they seemed so real, so informed, so photographically exact. It was like looking at something as it happened, rather than imagining it as it one day might be. I can remember studying with boundless fascination, and more than a touch of misplaced longing, a Bonestell illustration in *Life* magazine showing New York City at the moment of nuclear detonation, a giant mushroom cloud rising from the familiar landscape of central Manhattan, a second cloud spreading itself across the outlying sprawl of Queens. These illustrations were meant to frighten, but really they excited.*

*Bonestell was an interesting person. For most of his working life he was an architect, and ran a practice of national distinction in California until 1938 when, at the age of fifty, he abruptly quit his job and began working as a Hollywood film-set artist, creating background mattes for many popular movies. As a sideline he also began to illustrate magazine articles on space travel, creating imaginative views of moons and planets as they would appear to someone visiting from Earth. So when magazines in the Fifties needed lifelike illustrations of space stations and lunar launch pads, he was a natural and inspired choice. He died in 1986 aged ninety-eight.

I'm not suggesting that we actually wanted New York to be blown up – at least not exactly. I'm just saying that if it *did* ever happen you could see a plus side to it. We would all die, sure, but our last word would be a sincere and appreciative '*Wow*.'

Then in the late 1950s the Soviets briefly developed a clear lead in the space race and the excitement took on a real edge. The fear became that they would install giant space platforms in geostationary orbit directly above us, far beyond the reach of our gnat-like planes and weakly puffing guns, and that from this comfortable perch they would drop bombs on us whenever we peeved them.

In fact, that was never going to happen. Because of Earth's spin, you can't just drop bombs from space like water balloons. For one thing, they wouldn't drop; they would go into orbit. So you would have to fire them in some fashion, which required a level of delivery control the 1950s simply didn't command. And anyway because the Earth is spinning at a thousand kilometres an hour (give or take), you would have to master extremely precise trajectories to hit a given target. Any bomb fired from space was in fact far more likely to fall in a Kansas wheat field, or almost anywhere else on Earth, than through the roof of the White House. If bombarding each other from space had ever been a realistic option, we would have space stations up there in the hundreds now, believe me.

However, the only people who knew this in the

1950s were space scientists, and they weren't going to tell anybody because then we wouldn't give them money to develop their ambitious programmes. So magazines and Sunday supplements ran these breathless accounts of peril from above, because their reporters didn't know any better, or didn't wish to know any better, and because they had all these fantastic drawings by Chesley Bonestell that were such a pleasure to look at and just had to be seen.

So earthly devastation became both a constant threat and a happy preoccupation of that curiously bifurcated decade. Public service films showed us how private fallout shelters could be not only protective but *fun*, with Mom and Dad and Chip and Skip bunking down together underground, possibly for years on end. And why not? They had lots of dehydrated food and a whole stack of board games. 'And Mom and Dad need never worry about the lights running low with this handy pedal generator and two strong young volunteers to provide plenty of muscle power!' And no school! This was a lifestyle worth thinking about.

For those who didn't care to retreat underground, the Portland Cement Association offered a range of heavy-duty 'Houses for the Atomic Age!' – special 'all-concrete blast-resistant houses' designed to let the owners survive 'blast pressures expected at distances as close as 3,600 feet from ground zero of a bomb with an explosive force equivalent to 20,000 tons of TNT'. So the Russians could drop a bomb right in your own neighbourhood and you could sit in comfort at home reading the evening news-paper and hardly know there was a war on. Can you imagine erecting such a house and *not* wanting to see how well it withstood a nuclear challenge? Of course not. Let those suckers drop! We're *ready*!

And it wasn't just nuclear devastation that enthralled and excited us. The film world reminded us that we might equally be attacked by flying saucers or stiff-limbed aliens with metallic voices and deathly ray guns, and introduced us to the stimulating possibilities for mayhem inherent in giant mutated insects, blundering mega-crabs, bestirred dinosaurs, monsters from the deep, and one seriously pissed-off 50-foot woman. I don't imagine that many people, even those who now faithfully vote Republican, believed that any of that would actually happen, but certain parts of it – the UFOs and flying saucers, for instance – were far more plausible then than now. This was an age, don't forget, in which it was still widely believed that there might be civilizations on Mars or Venus. Almost anything was possible.

And even the more serious magazines like *Life* and *Look*, the *Saturday Evening Post*, *Time* and *Newsweek* found ample space for articles on interesting ways the world might end. There was almost no limit to what might go wrong, according to various theories. The Sun could blow up or abruptly wink out. We might be bathed in murder-ous radiation as Earth passed through the twinkly glitter of a comet's tail. We might have a new ice age. Or Earth might somehow become detached from its faithful orbit and drift out of the solar system, like a lost balloon, moving ever deeper into some cold, lightless corner of the universe. Much of the notion behind space travel was to get away from these irremediable risks and start up new lives with more interestingly padded shoulders inside some distant galactic dome.

Were people seriously worried about any of this? Who knows? Who knows what anyone in the 1950s was thinking about anything, or even if they were thinking at all. All I know is that any perusal of magazines from the period produces a curious blend of undiluted optimism and a kind of eager despair. Over 40 per cent of people in 1955 thought there would be a global disaster, probably in the form of world war, within five years and half of those were certain it would be the end of humanity. Yet the very people who claimed to expect death at any moment were at the same time busily buying new homes, digging swimming pools, investing in stocks and bonds and pension plans, and generally behaving like people who expect to live a long time. It was an impossible age to figure.

But even by the strange, elastic standards of the time, my parents were singularly unfathomable when it came to worry. As far as I can tell, they didn't fear a thing, even the things that other people really did worry about. Take polio. Polio had been a periodic feature of American life since the late 1800s (why it suddenly appeared then is a question that appears to have no answer) but it became particularly virulent in the early 1940s and remained at epidemic proportions well into the following decade, with between thirty thousand and forty thousand cases reported nationally every year. In Iowa, the worst year was 1952, which happened to be the first full year of my life, when there were over three thousand five hundred cases – roughly 10 per cent of the national total, or nearly three times Iowa's normal allotment – and one hundred and sixty-three deaths. A famous picture of the time from the *Des Moines Register* shows assorted families, including one man on a tall ladder, standing outside Blank Children's Hospital in Des Moines shouting greetings and encouragement to their quarantined children through the windows. Even after half a century it is a haunting picture, particularly for those who can remember just how unnerving polio was.

Several things made it so. First, nobody knew where it came from or how it spread. Epidemics mostly happened in the summer, so people associated polio with summer activities like picnics and swimming. That was why you weren't supposed to sit around in wet clothes or swallow pool water. (Polio was in fact spread through contaminated food and water, but swimming-pool water, being chlorinated, was actually one of the safer environments.) Second, it disproportionately affected young people, with symptoms that were vague and variable and always a worry to interpret. The best doctor in the world

283

couldn't tell in the initial stages whether a child had polio or just flu or a summer cold. For those who did get polio, the outcome was frighteningly unpredictable. Two-thirds of victims recovered fully after three or four days with no permanent ill effects at all. But others were partly or wholly paralysed. Some couldn't even breathe unaided. In the United States roughly 3 per cent of victims died; in outbreaks elsewhere it was as high as 30 per cent. Most of those poor parents calling through the windows at Blank Hospital didn't know which group their children would end up in. There wasn't a thing about it that wasn't a source of deepest anxiety.

Not surprisingly, a kind of panic came over communities when polio was reported. According to *Growing Up with Dick and Jane*, a history of the Fifties, at the first sign of a new outbreak, 'Children were kept away from crowded swimming pools, pulled out of movie theaters and whisked home from summer camps in the middle of the night. In newspapers and newsreels, images of children doomed to death, paralysis or years in an iron lung haunted the fearful nation. Children were terrified at the sight of flies and mosquitoes thought to carry the virus. Parents dreaded fevers and complaints of sore throats or stiff necks.'

Well, that's all news to me. I was completely unaware of any anxiety about polio. I knew that it existed – we had to line up from time to time after the mid-Fifties to get vaccinated against it – but I didn't know that we were supposed to be frightened. I didn't know about any dangers of any type anywhere. It was quite a wonderful position to be in really. I grew up in possibly the scariest period in American history and had no idea of it.

When I was seven and my sister was twelve, my dad bought a blue Rambler station wagon, a car so cruddy and styleless that even Edsel owners would slow down to laugh at you, and decided to break it in with a drive to New York. The car had no air conditioning, but my sister and I got the idea that if we laid the tailgate flat and stood on it, and held on to the roof rack, we could essentially get out of the car and catch a nice cooling breeze. In fact, it was like standing in the face of a typhoon. It couldn't have been more dangerous. If we relaxed our grip for an instant – to sneeze or satisfy an itch – we risked being whipped off our little platform and lofted into the face of a following Mack truck.

Conversely, if my father braked suddenly for any reason – and at least three or four times a day he provided us with sudden hold-on-to-your-hat swerves and a kind of bronco effect, braking when he dropped a lighted cigarette on to the seat between his legs and he and my mother jointly engaged in a frantic and generally entertaining search for it – there was a very good chance that we would be tossed sideways into a neighbouring field or launched – fired really – in a forward direction into the path of another mighty Mack.

It was, in short, insanely risky – a thought that evidently occurred to a highway patrolman near Ashtabula, Ohio, who set his red light spinning and pulled my dad over and chewed him out ferociously for twenty minutes for being so monumentally bone-headed with respect to his children's safety. My father took all this meekly. When the patrolman at last departed, my father told us in a quiet voice that we would have to stop riding like that until we crossed the state line into Pennsylvania in another hour or two.

It wasn't a terribly good trip for my dad. He had booked a hotel in New York from a classified ad in the *Saturday Review* because it was such a good deal, and then discovered that it was in Harlem. On the first night there, while my parents lay on the bed, exhausted from the ordeal of finding their way from Iowa to 1,252nd Street in upper Manhattan – a route not highlighted in any American Automobile Association guide – my sister and I decided to get something to eat. We strolled around the district for a while and found a corner diner about two blocks away. While we were sitting enjoying our hamburgers and chocolate sodas, and chatting amiably to several black people, a police car slid by, paused, backed up and pulled over. Two officers came in, looked around suspiciously, then came over to us. One of them asked us where we had come from.

'Des Moines, Iowa,' my sister replied.

'*Des Moines, Iowa!*' said the policeman, astounded. 'How did you get here from Des Moines, Iowa?'

'My parents drove us.'

'Your parents *drove* you here from Des Moines?'

My sister nodded.

'*Why?*'

'My dad thought it would be educational.'

'To come to Harlem?' The policemen looked at each other. 'Where are your parents now, honey?'

My sister told them that they were in the Hotel W. E. B. DuBois or Chateau Cotton Club or whatever it was.

'Your parents are staying *there*?'

My sister nodded.

'You really *are* from Iowa, aren'tcha, honey?'

The policemen took us back to the hotel and escorted us to our room. They banged on the door and my father answered. The policemen didn't know whether to be firm with my dad or gentle, to arrest him or give him some money or what. In the end they just strongly urged him to check out of the hotel first thing in the morning and to find a more appropriate hotel in a safer neighbourhood much lower down in Manhattan.

My father wasn't in a strong position to argue. For one thing, he was naked from the waist down. He was standing half behind the door so the police were unaware of his awkward position, but for those of us sitting on the bed the view was a memorably surreal one of my father, bare-buttocked, talking respectfully and in a grave tone of voice to two large New York policemen. It was a sight that I won't forget in a hurry.

My father was quite pale when the policemen left, and talked to my mother at length about what we were going to do. They decided to sleep on it. In the end, we stayed. Well, it was such a good rate, you see.

The *second* time I noticed that adults are not entirely to be trusted was also the first time I was genuinely made fearful by events in the wider world. It was in the autumn of 1962, just before my eleventh birthday, when I was home alone watching television and the programme was interrupted for a special announcement from the White House. President Kennedy came on looking grave and tired and indicated that things were not going terribly well with regard to the Cuban missile crisis – something about which at that point I knew practically nothing.

The background, if you need it, is that America had discovered that the Russians were preparing (or so we thought) to install nuclear weapons in Cuba, just ninety miles from American soil. Never mind that we had plenty of missiles aimed at Russia from similar distances in Europe. We were not used to being threatened in our own hemisphere and weren't going to stand for it now. Kennedy ordered Khrushchev to cease building launch pads in Cuba or else.

The Presidential address I saw was telling us that we were now at the 'or else' part of the scenario. I remember this as clearly as anything, largely because Kennedy looked worried and grey, not a look you wish to see in a

President when you are ten years old. We had installed a naval blockade around Cuba to express our displeasure and Kennedy announced now that a Soviet ship was on its way to challenge it. He said that he had given the order that if the Soviet ship tried to pass through the blockade, American destroyers were to fire in front of its bow as a warning. If it still proceeded, they were to sink it. Such an act would, of course, be the start of the Third World War. Even I could see that. This was the first time that my blood ever ran cold.

It was evident from Kennedy's tone that all this was pretty imminent. So I went and ate the last piece of a Toddle House chocolate pie that had been promised to my sister, then hung around on the back porch, wishing to be the first to tell my parents the news that we were all about to die. When they arrived home they told me not to worry, that everything would be all right, and they were right of course as always. We didn't die – though I came closer than anybody when my sister discovered that I had eaten her piece of pie.

In fact, we all came closer to dying than we realized. According to the memoirs of Robert McNamara, the then Secretary of Defense, the Joint Chiefs of Staff at that time suggested – indeed, eagerly urged – that we drop a couple of nuclear bombs on Cuba to show our earnest and to let the Soviets know that they had better not even think about putting nuclear weapons in our back yard. President Kennedy, according to McNamara, came very close to authorizing such a strike.

Twenty-nine years later, after the break-up of the Soviet Union, we learned that the CIA's evidence about Cuba was completely wrong (now there's a surprise) and that the Soviets in fact already had about one hundred and seventy nuclear missiles positioned on Cuban soil, all trained on us of course, and all of which would have been launched in immediate retaliation for an American attack. Imagine an America with one hundred and seventy of its largest cities – which, just for the record, would include Des Moines – wiped out. And of course it wouldn't have stopped there. That's how close we all came to dying.

I haven't trusted grown-ups for a single moment since.

Chapter 12
OUT AND ABOUT

JACKSON, MICH. (AP) – A teenaged girl and her 12-year-old brother were accused by police Saturday of trying to kill their parents by pouring gasoline on their bed and setting it alight while they were sleeping. The children told police their parents were 'too strict and were always nagging.' Mr. and Mrs. Sterling Baker were burned over 50 percent of their bodies and were listed in fair condition in a hospital.

– *Des Moines Tribune*, 13 June 1959

EVERY SUMMER, when school had been out for a while and your parents had had about as much of you as they cared to take in one season, there came a widely dreaded moment when they sent you to Riverview, a small, peeling amusement park in a dreary commercial district on the north side of town, with $2 in your pocket and instructions to enjoy yourself for at least eight hours, more if possible.

Riverview was an unnerving institution. The roller-coaster, a Himalayan massif of ageing wood, was the most rickety, confidence-sapping construction ever. The wagons were flocked inside and out with thirty-five years of spilled popcorn and hysterical vomit. It had been built in 1920, and you could feel its age in every groaning joint and cracked cross-brace. It was enormous – about four miles long, I believe, and some twelve thousand feet high. It was easily the scariest ride ever built. People didn't even scream on it; they were much too petrified to emit any kind of noise. As it passed, the ground would tremble with increasing intensity and it would shake loose a shower – actually a kind of avalanche – of dust and ancient birdshit from its filthy rafters. A moment later, there would be a passing rainshower of vomit.

The guys in charge of the rides were all closely modelled on Richard Speck, the Chicago murderer. They spent their working lives massaging zits and talking to groups of bouncy young women in bobby socks who unfathomably flocked to them. The rides weren't on timers of any kind, so if the attendants went off into their little booths to have sex, or fled over the fence and across the large expanse of open ground beyond at the appearance of two men with a warrant, the riders could be left on for indefinite periods – days if the employee had bolted with a vital key or crank. I knew a kid named Gus Mahoney who was kept on the Mad Mouse so long, and endured so many g-forces, that for three months afterwards he couldn't comb his hair forward and his ears almost touched at the back of his head.

Even the dodgem cars were insanely lively. From a distance the dodgem palace looked like a welder's yard because of all the sparks raining down from the ceiling, which always threatened to fall in the car with you, enlivening the ride further. The dodgem attendants didn't just permit head-on crashes, they actively encouraged them. The cars were so souped up that the instant you touched the accelerator, however lightly or tentatively, it would shoot off at such a speed that your head would become just a howling sphere on the end of a whip-like stalk. There was no controlling the cars once they were set in motion. They just flew around wildly, barely in contact with the floor, until they slammed into something solid, giving you the sudden opportunity to examine the steering wheel very closely with your face.

The worst outcome was to be caught in a car that turned out to be temperamental and sluggish or broke down altogether because forty other drivers, many of them small children who had never before had an opportunity to exact revenge on anything larger than a nervous toad, would fly into you with unbridled joy from every possible angle. I once saw a boy in a broken-down car bale out while the ride was still running – this was the one thing you *knew* you were never supposed to do – and stagger dazedly through the heavy traffic for the periphery. As he set foot on the metal floor, over two thousand crackling bluish strands of electricity leaped on to him from every direction, lighting him up like a paper lantern and turning him into a kind of living X-ray. You could see every bone in his body and most of his larger organs. Miraculously he managed to sidestep every car that came hurtling at him – and that was all of them, of course – and collapsed on the stubbly grass outside, where he lay smoking lightly from the top of his head and asked for someone to get word to his mom that he loved her. But apart from a permanent ringing in his ears, he suffered no major damage, though the hands on his Zorro watch were for ever frozen at ten past two.

There wasn't anything at Riverview that wasn't horrible. Even the Tunnel of Love was an ordeal. There was always a joker in the leading boat who would dredge up a viscous ball of phlegm and with a mighty *phwop* shoot it on to the low ceiling – an action that was known as hanging a louie. There it would dangle, a saliva stalactite, before draping itself over the face of a following boater. The trick of successful louie-hanging – and I speak here with some authority – had nothing to do with spit, but with how fast you could run when the boat stopped.

Riverview was where you also discovered that kids from the other side of town wanted you dead and were prepared to seize any opportunity in any dark corner to get you that way. Kids from the Riverview district went to a high school so forlorn and characterless that it didn't have a proper name, but just a geographical designation: North High. They detested kids from Theodore Roosevelt High School, the outpost of privilege, comfort and quality footwear for which we were destined. Wherever you went at Riverview, but particularly if you strayed from your group (or in the case of Milton Milton had no group), there was always a good chance that you would be pulled into the shadows and briskly drubbed and relieved of wallet, shoes, tickets and pants. There was always some kid – actually it was always Milton Milton now that I think of it – wandering in dismay in sagging undershorts or standing at the foot of the rollercoaster wailing helplessly at his jeans, now dangling from a rafter four hundred feet above the ground.

I knew kids who begged their parents not to leave them at Riverview, whose fingers had to be prised off car door handles and torn from any passing pair of adult legs, who left six-inch-deep grooves in the dust with their heels from where they were dragged from the car to the entrance gate and pushed through the turnstile, and told to have fun. It was like being put in a lion's cage.

The one amusement of the year that everyone did get genuinely excited about was the Iowa State Fair, which was held at enormous fairgrounds way out on the eastern edge of town late every August. It was one of the biggest fairs in the nation; the movie *State Fair* was filmed at and based on the Iowa State Fair, a fact that filled us all with a curious pride, even though no one to our knowledge had ever seen the movie or knew a thing about it.

The State Fair happened during the muggiest, steamiest period of the year. You spent all your time there soaked in perspiration and eating sickly foods – Sno Cones, cotton candy, ice cream bars, ice cream sandwiches, foot-long hot dogs swimming in gooey relish, bucketloads of the world's most sugary lemonade – until you had become essentially an ambulatory sheet of fly-paper and were covered from head to toe with vivid stains and stuck, half-dead insects.

The State Fair was mostly a celebration of the farming way of life. It had vast halls filled with quilts and jams and tasselled ears of corn and tables spread with dome-roofed pies the size of automobile tyres. Everything that could be grown, cooked, canned or sewn was carefully conveyed to Des Moines from every corner of the state and ardently competed over. There were also displays of shiny new tractors and other commercial manufactures in a hall of wonders known as the Varied Industries Building and every year there was something called the Butter Cow, which was a life-sized cow carved from an enormous (well, cow-sized) block of butter. It was considered one of the wonders of Iowa, and some way beyond, and always had an appreciative crowd around it.

Beyond the display buildings were ranks of enormous stinking pavilions, each several acres in size, filled with animal pens, mostly inhabited by hogs, and the amazing sight of hundreds of keen young men buffing, shampooing and grooming their beloved porkers in the hope of winning a coloured satin ribbon and bringing glory home to Grundy Center or Pisgah. It seemed an odd way to court fame.

For most people the real attraction of the fair was the midway with its noisy rides and games of chance and enticing sideshows. But there was one place that all boys dreamed of visiting above all others: the strippers' tent.

The strippers' tent had the brightest lights and most pulsating music. From time to time the barker would bring out some of the girls, chastely robed, and parade them around a little open-air stage while suggesting – and looking each of us straight in the eye – that these girls could conceive of no greater satisfaction in life than to share their natural bounties with an audience of appreciative, red-blooded young men. They all seemed to be amazingly good looking – but then I *was* running a temperature of over 113 degrees just from the thought of being on the same planet as young women of such miraculously obliging virtue, so I might have been a touch delirious.

The trouble was that we were twelve years old when we became seriously interested in the strippers' tent and you had to be thirteen to go in. A dangling sign on the ticket booth made that explicitly clear. Doug Willoughby's older brother, Joe, who was thirteen, went in and came out walking on air. He wouldn't say much other than that it was the best thirty-five cents he had ever spent. He was so taken that he went in three times more and pronounced it better on each occasion.

Naturally we circumnavigated the strippers' tent repeatedly looking for a breach of any kind, but it was the Fort Knox of canvas. Every millimetre of hem was staked to the ground, every metal eyelet sealed solid. You could hear music, you could hear voices, you could even see the shadowy outlines of the audience, but you couldn't discern the tiniest hint of a female form. Even Doug Willoughby, the most ingenious person I knew, was completely flummoxed. It was a torment to know that there was nothing but this rippling wall of canvas between us and living, breathing, unadorned female epidermis, but if Willoughby couldn't find a way through there wasn't a way through.

The following year I assembled every piece of ID I could find – school reports, birth certificate, library card, faded membership card from the Sky King Fan Club, anything that indicated my age even vaguely – and went directly to the tent with Buddy Doberman. It was newly painted with life-sized images of curvy pin-ups in the style of Alberto Vargas, and looked *very* promising.

'Two for the front row, please,' I said.

'Scram,' said the grizzled man who was selling tickets. 'No kids allowed.'

'Ah, but I'm thirteen,' I said, and began to extract affidavits from my folders.

'Not old enough,' said the man. 'You gotta be fourteen.' He hit the dangling sign. The '13' on it had been covered over with a square of card saying '14'.

'Since when?'

'Since this year.'

'But why?'

'New rules.'

'But that's not fair.'

'Kid, if you got a gripe, write to your congressman. I just take the money.'

'Yes, but . . .'

'You're holding up the line.'

'Yes, but . . .'

'Scram!'

So Buddy Doberman and I sloped off while a line of young men leered at us. 'Come back when you've all growed up,' yukked a young man from a place called, I would guess, Moronville, then vanished under a withering glance of ThunderGaze.

Getting into the strippers' tent would become the principal preoccupation of my pubescent years.

Most of the year we didn't have Riverview or the State Fair to divert us, so we went downtown and just fooled around. We were extremely good at just fooling around. Saturday mornings were primarily devoted to attaining an elevated position – the roofs of office buildings, the windows at the ends of long corridors in the big hotels – and dropping soft or wet things on shoppers below. We spent many happy hours too roaming through the behind-the-scenes parts of department stores and office buildings, looking in broom cupboards and stationery cabinets, experimenting with steamy valves in boiler rooms, poking through boxes in storerooms.

The trick was never to behave furtively, but to act as if you didn't realize you were in the wrong place. If you encountered an adult, you could escape arrest or detention by immediately asking a dumb question: 'Excuse me, mister, is this the way to Dr Mackenzie's office?' or 'Can you tell me where the men's room is, please?' This approach never failed. With a happy chuckle the apprehending custodian would guide us back to daylight and set us on our way with a pat on the head, unaware that under our jackets were thirteen rolls of duct tape, two small fire extinguishers, an adding machine, one semi-pornographic calendar from his office wall and a really lethal staple gun.

On Saturdays there were usually matinees to go to, generally involving a double feature of all the movies that my mother didn't take me to – *The Man from Planet X*, *Revenge of Godzilla*, *Zombies from the Stratosphere*, something with the slogan 'Half-man, half-beast, but ALL MONSTER' – plus a handful of cartoons and a couple of Three Stooges shorts just to make sure we were maximally fired up. The main features usually involved some fractious, jerkily animated dinosaurs, a swarm of giant mutated insects and several thousand severely worried Japanese people racing through city streets just ahead of a large wave or trampling foot.

These movies were nearly always cheaply made, badly acted and largely incoherent, but that didn't matter because Saturday matinees weren't about watching movies. They were about racing around wildly, making noise, having pitched battles involving thrown candy and making sure that every horizontal surface was buried at least three inches deep in spilled popcorn and empty containers. Essentially matinees were an invitation to four thousand children to riot for four hours in a large darkened space.

Before every performance, the manager – who was

nearly always a bad-tempered bald guy with a bow tie and a very red face – would take to the stage to announce in a threatening manner that if *any* child – any child at all – was caught throwing candy, or seemed to be about to throw candy, he would be seized by the collar and frog-marched into the waiting arms of the police. 'I'm watching you all, and I know where you live,' the manager would say and fix us with a final threatening scowl. Then the lights would dim and up to twenty thousand pieces of flying candy would rain down on him and the stage around him.

Sometimes the movies would be so popular or the manager so unseasoned and naive that the balcony would be opened, giving a thousand or so kids the joyous privilege of being able to tip wet and sticky substances on to the helpless swarms below. The running of the Paramount Theater was once entrusted to a tragically pleasant young man who had never dealt with children in a professional capacity before. He introduced an inter-mission in which children with birthdays who had filled out a card were called up on stage and allowed to reach into a big box from which they could extract a toy, box of candy or gift certificate. By the second week eleven thousand children had filled out birthday cards. Many were making seven or eight extra trips to the stage under lightly assumed identities. Both the manager and the free gifts were gone by the third week.

But even when properly run, matinees made no

economic sense. Every kid spent 35 cents to get in and another 35 cents on pop and candy, but left behind $4.25 in costs for repairs, cleaning and gum removal. In consequence matinees tended to move around from theatre to theatre – from the Varsity to the Orpheum to the Holiday to the Hiland – as managers abandoned the practice, had nervous breakdowns or left town.

Very occasionally the film studios or a sponsor would give out door prizes. These were nearly always ill-advised. For the premiere of *The Birds*, the Orpheum handed out one-pound bags of birdseed to the first five hundred customers. Can you imagine giving birdseed to five hundred unsupervised children who are about to go into a darkened auditorium? A little-known fact about birdseed is that when soaked in Coca-Cola and expelled through a straw it can travel up to two hundred feet at speeds approaching Mach 1 and will stick like glue to anything – walls, ceilings, cinema screens, soft fabrics, screaming usherettes, the back of the manager's suit and head, anything.

Because the movies were so bad, and the real action was out in the lobbies, nobody ever sat still for long. Once every half-hour or so, or sooner if nobody on the screen was staggering around with a stake through the eye or an axe in the back of his head, you would get up and go off to see if there was anything worth investigating in the theatre's public areas. In addition to the concession stands in the lobby, most theatres also had vending machines in dark, unsupervised corners, and these were always worth a look. There was a general conviction that just above where the cups dropped down or the candy bars slid out – slightly out of reach but tantalizingly close by – were various small levers and switches that would, if activated, dispense all the candy at once or possibly excite the change-release mechanism into setting loose a cascade of silvery coins. Doug Willoughby once brought a small flashlight and one of those angled mirrors that dentists use, and had a good look around the insides of a vending machine at the Orpheum, and became convinced that if he found someone with sufficiently long arms he could make the machine his servant.

So you may imagine the delight on his face on the day that someone brought him a kid who was about seven feet tall and weighed forty pounds. He had arms like garden hoses. Best of all, he was dim and pliant. Encouraged by a clutch of onlookers that quickly grew to a crowd of about two hundred, the kid dutifully knelt down and stuck his arm up the machine, probing around as Willoughby directed. 'Now go left a little,' Willoughby would say, 'past the capacitor, under the solenoid and see if you can't find a hinged lid. That'll be the change box. Do you feel it?'

'No,' the kid responded, so Willoughby fed in a little more arm.

'Do you feel it now?' Willoughby asked.

'No, but— ow!' the kid said suddenly. 'I just got a big shock.'

'That'll be the earthing clamp,' said Willoughby. 'Don't touch that again. I mean, really, don't touch that again. Try going around it.' He fed in a little more arm. 'Now do you feel it?'

'I can't feel anything, my arm's asleep,' said the kid after a time, and then added: 'I'm stuck. I think my sleeve's caught on something.' He grimaced and manoeuvred his arm, but it wouldn't come free. 'No, I'm really stuck,' he announced at last.

Somebody went and got the manager. He came bustling up a minute or so later accompanied by one of his oafish assistants.

'What the hell,' he growled, forcing his way through the crowd. 'Move aside, move aside. Goddam it all. What the hell. What the hell's going on? Goddam kids. *Move*, boy! Goddam it to hell. God damn. God damn. What the hell.' He reached the front of the crowd and saw, to his astonishment and disgust, a boy obscenely violating the innards of one of his vending machines. 'The hell you *doing*, buster? Get your arm out of there.'

'I can't. I'm stuck.'

The manager yanked on the kid's arm. The kid wailed in pain.

'Who put you up to this?'

'They all did.'

'Are you aware that it is a federal offence to tamper with the insides of a Food-O-Mat machine?' the manager said as he yanked more and the kid wailed. 'You are in a *world* of trouble, young man. I am going to personally escort you to the police station. I don't even want to *think* about how long you'll be in reform school, but you'll be shaving by the time of your next matinee, buster.'

The kid's arm would not come free, though it was now several inches longer than it had been earlier. Clucking, the manager produced an enormous ring of keys – the kind of ring that, once seen, made a man like him decide to drop all other plans and go straight into movie theatre management – unlocked the machine and hauled open the door, dragging the protesting kid along with it. For the first time in history the inside of a vending machine was exposed to children's view. Willoughby whipped out a pencil and notebook and began sketching. It was an entrancing sight – two hundred candy bars stacked in columns, each inhabiting a little tilted slot.

As the manager bent over to try to disentangle the kid's arm and shirt from the door, two hundred hands reached past him and deftly emptied the machine of its contents.

'Hey!' said the manager when he realized what was happening. Furious and sputtering, he snatched a large box of Milk Duds from a small boy walking past.

'Hey! That's mine!' protested the boy, grabbing back and holding on to the box with both hands. 'It's mine! I paid for it!' he shouted, feet flailing six inches off the floor. As they struggled, the box ripped apart and all the contents spilled out. At this, the boy covered his face with his hands and began weeping. Two hundred voices shrilly berated the manager, pointing out that the Food-O-Mat machine didn't dispense Milk Duds. During this momentary distraction the kid with the long arms slid out of his shirt and fled topless back into the theatre – an act of startling initiative that left everyone gaping in admiration.

The manager turned to his oafish assistant. 'Go get that kid and bring him to my office.'

The assistant hesitated. 'But I don't know what he looks like,' he said.

'Pardon?'

'I didn't see his face.'

'He's got no shirt on, you moron. He's bare-chested.'

'Yeah, but I still don't know what he looks like,' the assistant muttered, and stalked into the theatre, flashlight darting.

The boy with the long arms was never seen again. Two hundred kids had free candy. Willoughby got to study the inside of the vending machine and work out how it functioned. It was a rare victory for the inhabitants of Kid World over the dark, repressive forces of Adult World. It was also the last time the Orpheum ever had a children's matinee.

Willoughby was the smartest person I ever met, particularly with regard to anything mechanical or scientific. Afterwards he showed me the sketch he'd made when the door was open. 'It's astoundingly simple,' he said. 'I could hardly believe the lack of complexity. Do you know, it doesn't have an internal baffle or backflow gate or anything. Can you believe that?'

I indicated that I was prepared to be as amazed as the next man.

'There's nothing to stop reverse entry – nothing,' he said, shaking his head in wonder, and slid the plans into his back pocket.

The following week there was no matinee but we went to see *How the West Was Won*. About half an hour into the movie, he took me to the Food-O-Mat machine, reached into his jacket and pulled out two telescopic car aerials. Extending them, he inserted them into the machine, briefly manipulated them and down came a box of Dots.

'What would you like?' he said.

'Could I have some Red Hots?' I asked. I loved Red Hots.

He wriggled again and a box of Red Hots came down. And with that Willoughby became my best friend.

Willoughby was amazingly brainy. He was the first person I knew who agreed with me about Bizarro World, the place where things went backwards, though for rather more refined reasons than mine.

'It's preposterous,' he would agree. 'Think what it would do to mathematics. You couldn't have prime numbers any more.'

I'd nod cautiously. 'And when they got sick they'd have to suck puke back into their mouths,' I'd add, trying to get the conversation back to more comfortable territory.

'Geometry would be right out the window,' Willoughby would go on, and begin listing all the theorems that would fall apart in a world running in reverse.

We often had conversations like that, where we were both talking about the same thing, but from perspectives miles apart. Still it was better than trying to discuss Bizarro World with Buddy Doberman, who was surprised to learn it wasn't a real place.

Willoughby had an absolute genius for figuring out how to get fun out of unpromising circumstances. Once his dad came to give us a ride home from the movies, but told us that he had to stop at City Hall to pay his property taxes or something, so we were left sitting in the car at a meter outside an office building on Cherry Street for twenty minutes. Now normally this would be about as unpromising a circumstance as one could find oneself in, but as soon as his dad was round the corner, Willoughby bobbed out of the car and rotated the windscreen washer – I didn't even know you could do such a thing – so that it pointed towards the sidewalk, then got into the driver's seat and told me on no account to make eye contact with or seem to notice anyone passing by. Then each time someone walked past he would squirt them – and car windscreen washers put out a *lot* of water, a surprising amount, believe me.

The victims would stop in dumbfounded puzzlement on the spot where they had been drenched and look suspiciously in our direction – but we had the windows up and seemed completely oblivious of them. So they would turn to study the building behind them, and Willoughby would drill them in the back with another soaking blast. It was wonderful, the most fun I had ever had. I would be there still if it were up to me. Who would ever think to investigate a car windscreen washer for purposes of amusement?

Like me, Willoughby was a devotee of Bishop's, but he was a more daring and imaginative diner than I could ever be. He liked to turn on the table light and send the waitresses off on strange quests.

'Could I have some Angostura bitters, please?' he would say with a look of choirboy sweetness. Or: 'Please could I have some fresh ice cubes; these are rather misshapen.' Or: 'Would you by any chance have a spare ladle and some tongs?' And the waitresses would go clumping off to see what they could find for him. There was something about his cheery face that inspired an eagerness to please.

On another occasion he pulled from his pocket, with a certain theatrical flourish, a neatly folded white handkerchief from which he produced a perfectly preserved large, black, flat, ugly, pincered stag beetle – what was known in Iowa as a June bug – and set it adrift on his tomato soup. It floated beautifully. One might almost have supposed it had been designed for the purpose.

Then he put the table light on. An approaching waitress, spying the beetle, shrieked and dropped an empty tray, and got the manager, who came hastening over. The manager was one of those people who are so permanently and comprehensively stressed that even their hair and clothes appear to be at their wits' end. He looked as if he had just stepped from a wind tunnel. Seeing the floating insect, he immediately embarked on a nervous breakdown.

'Oh my goodness,' he said. 'Oh my goodness, my goodness. I don't know how this has happened. This has never happened before. Oh my goodness, I am so sorry.' He whisked the offending bowl off the table, holding it at arm's length, as if it were actively infectious. To the waitress he said, 'Mildred, get these young men whatever they want – what*ever* they want.' To us he said: 'How about a couple of hot fudge sundaes? Would that help to fix matters for you?'

'Yes, please!' we replied.

He snapped his fingers and sent Mildred off to get us sundaes. 'With plenty of nuts and extra cherries,' he called. 'And don't forget the whipped cream.' He turned to us more confidentially. 'Now you won't tell anyone about this, will you, boys?' he said.

We promised not to.

'What do your parents do?'

'My father's a health inspector,' Willoughby said brightly.

'Oh my God,' said the manager, draining of blood, and rushed off to make sure our sundaes were the largest and most elaborate ever served at Bishop's.

The following Saturday, Willoughby led me into Bishop's again. This time he drank half his water, then pulled from his jacket a jar filled with pondwater, which he used to top up his glass. When he held the glass up to the light there were about sixteen tadpoles swimming in it.

'Excuse me, should my water be like this?' he called to a passing waitress, who stared at the water with a transfixed look, then went off to get back-up. Within a minute we had half a dozen waitresses examining the water with consternation, but no shrieking. A moment later our friend the manager turned up.

He held the glass up. 'Oh my *good*ness,' he said and went pale. 'I am so sorry. I don't know how this could have happened. Nothing like this has ever happened before.' He looked at Willoughby more closely. 'Say, weren't you here last week?'

Willoughby nodded apologetically.

I assumed we were about to be heaved out on our ears, but the manager said: 'Well, I am so sorry again, son. I cannot apologize enough.' He turned to the waitresses. 'This young man seems to be jinxed.' To us, he said, 'I'll get your sundaes,' and went off to the kitchen, pausing here and there en route to crouch down and look discreetly at the water of other diners.

The one thing Willoughby always lacked was a sense of proportion. I begged him not to push his luck, but the following week he insisted on going to Bishop's again. I refused to sit with him, but took a table across the way and watched as he hummingly pulled from his pocket a brown paper bag and carefully tipped into his soup about two pounds of dead flies and moths that he had retrieved from the overhead light fitting in his bedroom. They formed a mound about four inches high. It was a magnificent sight, but perhaps just a touch deficient in terms of plausibility.

By chance the manager was passing as Willoughby put on his light. The manager looked at the offending bowl in horror and utter dismay and then at Willoughby. I thought for a moment that he was going to faint or perhaps even die. 'This is just not poss—' he said and then a giant light bulb went on over his head as he realized that indeed it wasn't possible for anyone to be served a bowl of soup with two pounds of dead insects in it.

With commendable restraint he escorted Willoughby to the street door, and asked him – not demanded, but just asked him quietly, politely, sincerely – never to return. It was a terrible banishment.

All the Willoughbys – mother, father, four boys – were touched with brilliance. I used to think we had a lot of books in our house because of our two big bookcases in the living room. Then I went to the Willoughbys' house. They had books and bookcases *everywhere* – in the hallways and stairwells, in the bathroom, the kitchen, around all the walls of the living room. Moreover, theirs were works of real weight – Russian novels, books of history and philosophy, books in French. I realized then we were hopelessly outclassed.

And their books were read. I remember once Willoughby showed me a paragraph about farmboy bestiality he had come across in a long article about something else altogether in the *Encyclopaedia Britannica*. I don't remember the details now – it's not the sort of thing one retains for forty years – but the gist of the passage was that 32 per cent of farmboys in Indiana (or something like that: I'm pretty sure it was Indiana; it was certainly a high number) at one time or another had enjoyed sexual congress with livestock.

This amazed me in every possible way. It had never occurred to me that any farmboy or other human being, in Indiana or elsewhere, would ever willingly have sex with an animal, and yet here was printed evidence in a respectable publication that a significant proportion of them had at least given it a try. (The article was a touch coy on how enduring these relationships were.) But even more amazing than the fact itself was the finding of it. The *Encyclopaedia Britannica* ran to twenty-three volumes spread over eighteen thousand pages – some fifty million words in all, I would estimate – and Willoughby had found the only riveting paragraph in the whole lot. How did he do it? Who *reads* the *Encyclopaedia Britannica*?

Willoughby and his brothers opened new worlds, unsuspected levels of possibility, for me. It was as if I had wasted every moment of existence up to then. In their house anything could be fascinating and entertaining. Willoughby shared a bedroom with his brother Joe, who was one year older and no less brilliant at science. Their room was more laboratory than bedroom. There was apparatus everywhere – beakers, vials, retorts, Bunsen burners, jars of chemicals of every description – and books on every subject imaginable, all well-thumbed: applied mechanics, wave mechanics, electrical engineering, mathematics, pathology, military history. The Willoughby boys were always doing something large scale and ambitious. They made their own helium balloons. They made their own rockets. They made their own gunpowder. One day I arrived to find that they had built a rudimentary cannon – a test model – out of a piece of metal pipe into which they stuffed gunpowder, wadding and a silver ball-bearing about the size of a marble. This they laid on an old tree stump in their back yard, aimed at a sheet of plywood about fifteen feet away. Then they lit the fuse and we all retired to a safe position behind a picnic table turned on its side (in case the whole thing blew up). As we watched, the burning fuse somehow unbalanced the pipe and it began to roll slowly across the stump, taking up a new angle. Before we could react, it went off with a stupendous bang and blew out an upstairs bathroom window of a house three doors away. No one was hurt, but Willoughby was grounded for a month – he was commonly grounded – and had to pay $65 restitution.

The Willoughby boys really were able to make fun out of nothing at all. On my first visit, they introduced me to the exciting sport of match fighting. In this game, the competitors arm themselves with boxes of kitchen matches, retire to the basement, turn off all the lights and spend the rest of the evening throwing lighted matches at each other in the dark.

In those days kitchen matches were heavy-duty implements – more like signal flares than the weedy sticks we get today. You could strike them on any hard surface and fling them at least fifteen feet and they wouldn't go out. Indeed, even when being beaten vigorously with two hands, as when lodged on the front of one's sweater, they seemed positively determined not to fail. The idea, in any case, was to get matches to land on your opponents and create small, alarming bush fires on some part of their person; the hair was an especially favoured target. The drawback was that each time you launched a lighted match you betrayed your own position to anyone skulking in the dark nearby, so that after an attack on others you were more or less certain to discover that your own shoulder was robustly ablaze or that the centre of your head was a beacon of flame fuelled from a swiftly diminishing stock of hair.

We played for three hours one evening, then turned on the lights and discovered that we had all acquired several amusing bald patches. Then we walked in high spirits down to the Dairy Queen on Ingersoll Avenue for refreshment and a breath of air, and came back to discover two fire engines out front and Mr Willoughby in an extremely animated state. Apparently we had left a match burning in a laundry basket and it had erupted in flames, climbed up the back wall and scorched a few rafters, filling much of the house above with smoke. To all of this a team of firemen had enthusiastically added a great deal of water, much of which was now running out the back door.

'What were you *doing* down there?' Mr Willoughby asked in amazement and despair. 'There must have been eight hundred spent matches on the floor. The fire marshal is threatening to arrest me for arson. In my own house. What were you *doing*?'

Willoughby was grounded for six weeks after that, and so we had to suspend our friendship temporarily. But that was OK because by chance I had also become friends at this time with another schoolmate named Jed Mattes, who offered a complete contrast to Willoughby. For one thing, Jed was gay, or at least soon would be.

Jed had charm and taste and impeccable manners, and thanks to him I was exposed to a more refined side of life – to travel, quality food, literary fiction, interior design. It was unexpectedly refreshing. Jed's grandmother lived in the Commodore Hotel on Grand Avenue, which was rather an exotic thing to do. She was over a thousand years old and weighed thirty-seven pounds, which included sixteen pounds of make-up. She used to give us money to go to the movies, sometimes quite enormous sums, like $40 or $50, which would buy you a very nice day out in the early 1960s. Jed never wanted to go to movies like *Attack of the 50-Foot Woman*. He favoured musicals like *The Unsinkable Molly Brown* or *My Fair Lady*. I can't say these were my absolute first choices, but I went with him in a spirit of friendship and they did lend me a certain sheen of cosmopolitanism. Afterwards, he would take us in a cab – to me a form of conveyance of impossible elegance and splendour – to Noah's Ark, an esteemed Italian eatery on Ingersoll. There he introduced me to spaghetti and meatballs, garlic bread, and other worldly dishes of a most sophisticated nature. It was the first time I had ever been presented with a linen napkin or been confronted with a menu that wasn't laminated and slightly sticky and didn't have photographs of the food in it.

Jed could talk his way into anything. We used often to go and look in the windows of rich people's houses. Occasionally he would ring the front door bell.

'Excuse me for intruding,' he would say when the lady of the house arrived, 'but I was just admiring your living room curtains and I simply have to ask, where did you find that velour? It's *won*derful.'

319

The next thing you knew we'd be in the house, getting a full tour, with Jed cooing in admiration at the owner's inspired improvements and suggesting modest additional touches that might make it better still. By such means we became welcome in all the finest homes. Jed struck up a particular friendship with an aged philanthropist named A. H. Blank, founder of Blank Children's Hospital, who lived with his tottering, blue-haired wife in a penthouse apartment in the ritziest and most fashionable new address in Iowa, a building called The Towers, on Grand Avenue. Mr and Mrs Blank owned the whole of the tenth floor. It was the highest apartment between Chicago and Denver, or at the very least Grinnell and Council Bluffs, they told us. On Friday nights we would often stop by for cocoa and coffee cake and a view of the city – indeed of most of the Midwest, it seemed – from the Blanks' extensive balconies. It was in every sense the high point of all our weeks. I waited years for Mr Blank to die in the hope that he would leave me something, but it all went to charity.

One Saturday after going to the movies (*Midnight Lace* starring Doris Day, which we immediately agreed was OK but by no means one of her best), we were walking home along High Street – an unusual route; a route for people of an adventurous disposition – when we passed a small brick office building with a plaque that said 'Mid-America Film Distribution' or something like that, and Jed suggested we go in.

Inside, a small, elderly man in a lively suit was sitting at a desk doing nothing.

'Hello,' said Jed, 'I hope I'm not intruding, but do you have any old film posters you don't require any longer?'

'You like movies?' said the man.

'Like them? Sir, no, I *love* them.'

'No kidding,' said the man, pleased as anything. 'That's great, that's great. Tell me, son, what's your favourite movie?'

'I think that would have to be *All About Eve*.'

'You like that?' said the man. 'I've got that here somewhere. Hold on.' He took us into a storeroom that was packed from floor to ceiling with rolled posters and began searching through them. 'It's here somewhere. What else you like?'

'Oh, gosh,' said Jed, '*Sunset Boulevard*, *Rebecca*, *An Affair to Remember*, *Lost Horizon*, *Blithe Spirit*, *Adam's Rib*, *Mrs Miniver*, *Mildred Pierce*, *The Philadelphia Story*, *The Man Who Came to Dinner*, *Now Voyager*, *A Tree Grows in Brooklyn*, *Storm Warning*, *The Pajama Game*, *This Property Is Condemned*, *The Asphalt Jungle*, *The Seven-Year Itch*, *From This Day Forward* and *How Green Was My Valley*, but not necessarily in that order.'

'I got those!' said the man excitedly. 'I got all those.' He started passing posters to Jed in a manic fashion. He turned to me. 'What about you?' he said politely.

'*The Brain That Wouldn't Die*,' I answered hopefully.

He grimaced and shook his head. 'I don't handle B stuff,' he said.

'*Zombies on Broadway*?'

He shook his head.

'*Island of the Undead*?'

He gave up on me and turned back to Jed. 'You like Lana Turner movies?'

'Of course. Who doesn't?'

'I've got 'em all – every one since *Dancing Co-Ed*. Here, I want you to have them.' And he began piling them on to Jed's arms.

In the end, he gave us more or less everything he had – posters dating back to the late 1930s, all in mint condition. Goodness knows what they would be worth now. We took them in a cab back to Jed's house and divided them up on his bedroom floor. Jed took all the ones for movies starring Doris Day and Debbie Reynolds. I got the ones with men running along in a crouch with guns blazing. We were both extremely happy.

Some years later, I went away to Europe for a summer and ended up staying two years. While I was away my parents cleared out my bedroom. The posters went on a bonfire.

There were certain things I couldn't comfortably share with Jed and the one that stuck out most was my lustful wish to see a naked woman. I don't think an hour passed in the 364 days following my rejection at the State

Fairgrounds that I didn't think at least twice about the strippers' tent. It was the only possible place to see naked female flesh *in* the flesh, and my need was growing urgent.

By the March following my fourteenth birthday, I was crossing off on a calendar the number of days till the State Fair. By late June I was frequently short of breath. On 20 July I laid out the clothes I was going to wear the following month. It took me three hours to choose. I considered taking opera glasses, but decided against it on the grounds that they would probably steam up.

The official opening of the fair was 20 August. Normally no sane person went to the State Fair on its opening day because the crowds were so vast and suffocating, but Doug Willoughby and I went. We had to. We just had to. We met soon after dawn and took a bus all the way out to the east side. There we joined the cheerful throngs and waited three hours in line to be among the first in.

At 10 a.m. the gates swung open and twenty thousand people went whooping across the landscape, like the attacking hordes in *Braveheart*. You may be surprised to hear that Willoughby and I didn't go straight to the strippers' tent but rather bided our time. It was our considered intention to savour the occasion, so we had a good look round the exhibition halls. Possibly this was the first time in history that anyone has treated quilts and a butter cow as a form of foreplay, but we knew what we were doing. We wanted to let the girls have a chance to limber up, get into their stride. We didn't wish to attend an inferior show on our first visit.

At 11 a.m. we fortified ourselves with a popular ice cream confection known as a Wonder Bar, then proceeded to the strippers' tent and took our place in the line, pleased to be taking up one of the privileges of our seniority. But shortly before reaching the ticket booth, Willoughby nudged me in the ribs and indicated the dangling sign. It was new and it said: 'Absolutely NO MINORS! You must be SIXTEEN and have GENUINE I.D.'

I was speechless. At this rate, I would be getting a senior citizen discount by the time I saw my first naked woman.

At the window the man asked how old we were.

'Sixteen,' said Willoughby briskly, as if he would say anything else.

'You don't look sixteen to me, kid,' said the man.

'Well, I have a slight hormone deficiency.'

'You got I.D.?'

'No, but my friend here will vouch for me.'

'Fuck off.'

'But we were rather counting on attending one of the shows, you see.'

'Fuck off.'

'We've been waiting for this day for a year. We've been here since six a.m.'

'Fuck off.'

And so we slunk away. It was the cruellest blow I had suffered in my life.

The following week I went to the fair with Jed. It was an interesting contrast since he spent hours in the farmwives' section chatting to ladies in frilly-edged aprons about their jams and quilts. There wasn't a thing in the world of domestic science that didn't fascinate him and not a single obstacle or potential setback that didn't awake his immediate compassion. At one point he had a dozen women, all looking like Aunt Bea on the *Andy Griffith Show*, gathered round, all enjoying themselves immensely.

'Well, wasn't that just *wonderful*?' he said to me afterwards and gave an enormous happy sigh. 'Thank you so much for indulging me. Now let's take you to the strippers' tent.'

I had told him about my disappointment the previous week, and reminded him now that we were too young to gain admission.

'Age is but a technicality,' he said breezily.

At the tent, I held back while Jed went up to the ticket window. He talked to the man for some time. Occasionally they both looked at me, nodding gravely, as if in agreement about some notable deficiency on my part. Eventually Jed came back smiling and handed me a ticket.

'There you go,' he said cheerfully. 'I hope you don't mind if I don't join you.'

I was quite unable to speak. I looked at him in wonder and with difficulty stammered: 'But how?'

'I told him you had an inoperable brain tumour, which he didn't quite buy, and then I gave him ten bucks,' Jed explained. 'Enjoy!'

Well, what can I say other than that it was the highlight of my life? The stripper – there was only one per show, it turned out, something Willoughby's brother had neglected to tell us – was majestically bored, sensationally bored, but there was something unexpectedly erotic in her pouty indifference and glazed stare, and she really wasn't bad looking. She didn't strip off completely. She retained a sequined blue G-string and had nipple caps and tassels on her breasts, but it was still a divine experience, and when, as a kind of climax – a term I use advisedly but with a certain scientific precision – she leaned out over the audience, not six feet from my adoring gaze, and gave a ten-second twirl of the tassels, propelling them briefly but expertly in *opposite* directions – what a talent was this! – I thought I had died and that this was heaven.

I still firmly believe it will be much like that if I ever get there. And knowing that, there has scarcely been a moment in all the years since that I have not been extremely good.

Chapter 13
THE PUBIC YEARS

In Coeur D'Alene, Idaho, after householders reported that a car was tearing around the neighborhood in reverse, Assistant Police Chief Robert Schmidt investigated and found behind the wheel a teen-age girl who explained: 'My folks let me have the car, and I ran up too much mileage. I was just unwinding some of it.'

– *Time* magazine, 9 July 1956

ACCORDING TO THE Gallup organization 1957 was the happiest year ever recorded in the United States of America. I don't know that anyone has ever worked out why that largely uneventful year should have marked the giddy peak of American bliss, but I suspect it is more than coincidental that the very next year was the year that the New York Giants and the Brooklyn Dodgers dumped their hometown fans and decamped to California.

Goodness knows it was time for baseball to expand westward – it was ridiculous to have teams crammed into the old cities of the East and Midwest but not in any of the newer municipal colossi of the Western states – but the owners of the Dodgers and Giants weren't doing it for the good of baseball. They were doing it out of greed. We were entering a world where things were done because they offered a better return, not a better world.

People were wealthier than ever before, but life somehow didn't seem as much fun. The economy had become an unstoppable machine: gross national product rose by 40 per cent in the decade, from about $350 billion in 1950 to nearly $500 billion ten years later, then rose by another third to $658 billion in the next six years. But what had once been utterly delightful was now becoming very slightly, rather strangely unfulfilling. People were beginning to discover that joyous consumerism is a world of diminishing returns.

By the closing years of the 1950s most people – certainly most middle-class people – had pretty much everything they had ever dreamed of, so increasingly there was nothing much to do with their wealth but buy more and bigger versions of things they didn't truly require: second cars, lawn tractors, double-width fridges, hi-fis with bigger speakers and more knobs to twiddle, extra phones and televisions, room intercoms, gas grills, kitchen gadgets, snowblowers, you name it. Having more things of course also meant having more complexity in one's life, more running costs, more things to look after, more things to clean, more things to break down. Women increasingly went out to work to help keep the whole enterprise afloat. Soon millions of people were caught in a spiral in which they worked harder and harder to buy labour-saving devices that they wouldn't have needed if they hadn't been working so hard in the first place.

By the 1960s, the average American was producing twice as much as only fifteen years before. In theory at least, people could now afford to work a four-hour day, or two-and-a-half-day week, or six-month year and still maintain a standard of living equivalent to that enjoyed by people in 1950 when life was already pretty good – and arguably, in terms of stress and distraction and sense of urgency, in many respects much better. Instead, and almost uniquely among developed nations, Americans took none of the productivity gains in additional leisure. We decided to work and buy and have instead.

Of course not everyone shared equally in the good times. Black people who tried to improve their lot, particularly in the Deep South, particularly in Mississippi, were often subjected to the most outrageous and shocking abuse (made all the more so by the fact that most people at the time didn't seem shocked or outraged at all). Clyde Kennard, a former Army sergeant and paratrooper and a person of wholly good character, tried to enrol at Mississippi Southern College in Hattiesburg in 1955. He was sent away, but thought it over and came back and asked again. For this repetitive wilful uppitiness, university officials – I'll just make that quite clear: not students, not under-educated townspeople in white sheets, but university officials – planted illicit liquor and a bag of stolen chicken feed in his car and had him charged with grand theft. Kennard was tried and sent to prison for seven years for crimes he didn't commit. He died there before his term was completed.

Elsewhere in Mississippi at that time the Reverend George Lee and a man named Lamar Smith tried, in separate incidents, to exercise their right to vote. Smith actually succeeded in casting a ballot – in itself something of a miracle – but was shot dead on the courthouse steps five minutes later as he emerged with a dangerously triumphant smile. Although the killing was in broad daylight in a public place, no witnesses came forward and no assailant was ever charged. The Reverend Lee, meanwhile, was turned away at his polling station, but shot dead anyway, with a shotgun from a passing car as he drove home that night. The Humphreys County sheriff ruled the death a traffic accident; the county coroner recorded it as being of unknown causes. There were no convictions in that case either.

Perhaps the most shocking episode of all occurred in Money, Mississippi, when a young visitor from Chicago named Emmett Till rashly whistled at a white woman outside a country store. That evening Till was hauled from his relatives' house by white men, driven to a lonesome spot, beaten to a pulp, shot dead and dumped in the Tallahatchie River. He was fourteen years old.

Because Till was so young and because his mother in Chicago insisted on leaving the coffin open so that the world could see what her son had suffered, there was, finally, a national outcry. In consequence, two men – the husband of the woman who had been whistled at and his half-brother – were arrested and a trial was duly held. The evidence against the two was pretty overwhelming. They hadn't done much to cover their tracks, but then they didn't need to. After less than an hour's deliberation, the jury – all local people, all white of course – found them not guilty. The verdict would have been quicker, remarked the grinning foreman, if the jurors hadn't taken a break to drink a bottle of pop. The next year, knowing that they could never be retried, the two accused men happily admitted in an interview in *Look* magazine that they had indeed beaten and killed young Till.

Meanwhile, things weren't going terribly well for America in the wider world. In the autumn of 1957, the Soviets successfully tested their first intercontinental ballistic missile, which meant that now they could kill us without leaving home, and within weeks of that they launched the world's first satellite into space. Called Sputnik, it was a small metal sphere about the size of a beachball that didn't do much but orbit the Earth and go 'ping' from time to time, but that was very considerably more than we could do. The following month the Soviets launched Sputnik II, which was much larger at eleven hundred pounds and carried a little dog (a little *Communist* dog) called Laika. Our vanity stung, we responded by announcing a satellite launch of our own, and on 6 December 1957, at Cape Canaveral in Florida, the burners were fired on a giant Viking rocket carrying a fancy new Vanguard satellite. As the world watched, the rocket slowly rose two feet, toppled over and exploded. It was a humiliating setback. The papers referred to the incident variously as 'Kaputnik', 'Stayputnik', 'Sputternik' or 'Flopnik', depending on how comfortable they were with wit. President Eisenhower's normally steady popularity ratings dropped twenty-two points in a week.

America didn't get its first satellite into space until

1958 and that wasn't awfully impressive: it weighed just thirty-one pounds and was not much larger than an orange. All four other major launches by the US that year crashed spectacularly or refused to take to the air. As late as 1961, over a third of US launches failed.

The Soviets meanwhile went from strength to strength. In 1959 they landed a rocket on the Moon and took the first pictures of its backside, and in 1961 successfully put the first astronaut, Yuri Gagarin, into space and safely brought him home again. One week after the Gagarin space trip came the disastrous American-led Bay of Pigs invasion in Cuba, bringing extra layers of embarrassment and worry to national life. We were beginning to look hopeless and outclassed at whatever we did.

News from the world of popular culture was generally discouraging as well. Research showed that cigarettes really did cause cancer, as many people had long suspected. Tareyton, my father's brand, quickly rushed out a series of ads calmly reassuring smokers that 'All the tars and nicotine trapped in the filter are guaranteed not to reach your throat' without mentioning that all the lethal goos *not* trapped in the filter would. But consumers weren't so easily taken in by fatuous and misleading claims any longer, particularly after news came out that advertisers had been engaged in secret trials of devious subliminal advertising. During a test at a movie house in Fort Lee, New Jersey, patrons were shown a film in which two clipped phrases – 'Drink Coca-Cola' and 'Hungry? Eat Popcorn' – were flashed on the screen for 1/3000 of a second every five seconds – much too fast to be consciously noted, but subconsciously influential, or so it seemed, for sales of Coke went up 57.7 per cent and popcorn by nearly 20 per cent during the period of the experiment, according to *Life* magazine. Soon, *Life* warned us, all movies and television programmes would be instructing us hundreds of times an hour what to eat, drink, smoke, wear and think, making consumer zombies of us all. (In fact, subliminal advertising didn't work and was soon abandoned.)

Elsewhere on the home front, juvenile crime continued to rise and the education system seemed to be falling apart. The most popular non-fiction book of 1957 was an attack on American education standards called *Why Johnny Can't Read*, warning us that we were falling dangerously behind the rest of the world, and linking the success of Communism to a decline in American reading. Television got itself into a terrible scandal when it was revealed that many of the game shows were rigged. Charles Van Doren, boyish, modest, good-looking scion of a family of distinguished academics and intellectuals (his father and uncle had both won Pulitzer prizes), became a national hero, held up as a model to youngsters for his good manners and lack of swagger, while winning almost $130,000 on the programme *Twenty-One*, but then had to admit that he had been fed the answers. So had many other contestants on other shows, including a Protestant minister named Charles Jackson. Wherever you looked, it was just one bad thing after another. And nearly all this disturbed tranquillity occurred in the space of just over a year. People have never gone from happy to not happy more quickly.

In Des Moines as the decade came to an end the change was mostly physical. Chain stores and restaurants began to come in, causing flurries of excitement wherever they arose. Now we would be able to dine at the same restaurants, eat the same fast foods, wear the same clothes, direct visitors to the same motel beds as people in California and New York and Florida. Des Moines would be exactly like everywhere else, a prospect that most people found rather thrilling.

The city lost its elm trees to Dutch elm disease, leaving the main thoroughfares looking starkly naked. Often now along streets like Grand and University avenues the old houses were bulldozed to splinters, and in no time at all there would rise in their place a bright new gas station, a glassy restaurant, an apartment complex in a sleek modern style, or just a roomy new parking lot for a neighbouring business. I remember going away one year on vacation (a tour of Pony Express routes of the Plains States) and coming home to find that two stately Victorian houses across from Tech High School on Grand had become sudden vague memories. In their place, in what now seemed an enormous clearing, stood a sun-catching, concrete-white, multi-storey Travelodge motel. My father was apoplectic, but most people were pleased and proud – the Travelodge was more than just a motel, you see: it was a *motor lodge*, something far finer; Des Moines was coming up in the world – and I was both amazed and impressed that such a dramatic change could be effected so quickly.

At about the same time, a Holiday Inn opened on Fleur Drive, a park-like boulevard, mostly residential, leading from the city to the airport. It was a comparatively discreet building, but it had an enormous, exceedingly lively sign by the roadside – a thrumming angular tower of starbursts and garish cascades and manic patterns made by light bulbs chasing after each other in tireless circles – that exercised my father greatly. 'How could they let them put up a sign like that?' he would despair every time we drove past it from 1959 to his death twenty-five years later. 'Have you ever seen anything more ugly in your life?' he would ask no one in particular.

I thought it was wonderful. I couldn't wait for more signs like it everywhere, and I quickly got my wish as newer, more insistent, more car-friendly businesses popped up all over. In 1959, Des Moines got its first shopping mall, way out on Merle Hay Road, a part of town so remote, so out in the fields, that many people had to ask where it was. The new mall had a parking lot the size of a New England state. No one had ever seen so much asphalt in one place. Even my father got excited by this.

'Wow, look at all the places you can *park*,' he said, as if for all these years he had been cruising endlessly, unable to terminate a journey. For about a year the most dangerous place to drive in Des Moines was the parking lot of Merle Hay Mall because of all the cars speeding at joyous random angles across its boundless blacktop without reflecting that other happy souls might be doing likewise.

My father never shopped anywhere else after that. Neither did most people. By the early 1960s, people exchanged boasts about how long it had been since they had been downtown. They had found a new kind of happiness at the malls. At just the point where I was finally growing up, Des Moines stopped feeling like the place I had grown up in.

After Greenwood I moved on to Callanan Junior High School for grades seven to nine – the early teen years. Callanan was a much worldlier school. Its catchment area covered a broader cross-section of the city so that its enrolment was roughly half black and half white. For many of us this was our first close-up experience of black kids. Suddenly there were six hundred fellow students who were stronger, fleeter, tougher, braver, hipper and cannier than we were. This was when you realized for certain something that you had always privately suspected – that you were never going to take Bob Cousy's place on the Boston Celtics, never going to break Lou Brock's base-stealing records for the St Louis Cardinals, never going to be invited to Olympic trials in any sport. You weren't even going to make the junior varsity softball team now.

This was evident from the very first day when Mr Schlubb, the pear-shaped PE teacher, sent us all out to run half a dozen laps round a preposterously enormous cinder track. For the Greenwood kids – all of us white, marshmallowy, innately unphysical, squinting unfamiliarly in the bright sunshine – it was a shock to the system of an unprecedented order. Most of us ran as if slogging through quicksand and were gasping for air by the first bend. On the second lap a boy named Willis Pomerantz burst into tears because he had never perspired before and thought he was leaking vital fluids, and three others petitioned to be sent to the nurse. The black kids without exception sailed past us in a jog, including a three-hundred-pound spheroid named Tubby Brown. These kids weren't just slightly better than us, they were better by another order of magnitude altogether, and it was like this, we would find, in all sports.

Winters at Callanan were spent playing basketball in a dim-lit gymnasium – we seemed to spend hours at it every day – and no white kid I know ever even *saw* the ball. Honestly. You would just see a sequence of effortless blurs moving about between two or three lanky black kids and then the net would go swish, and you would know to turn round and lope down to the other end of the court.

Mostly you just tried to stay out of the way, and never ever raised your hands above your waist, for that might be taken as a sign that you wanted a pass, which was in fact the last thing you wanted. A boy named Walter Haskins once unthinkingly scratched the side of his head near the basket, and the next instant was hit square in the face so hard with a ball that the front of his head went completely concave. They had to use a bathroom plunger to get it back to normal, or so I was told.

The black kids were all immensely tough, too. I once saw an overfed white lummox named Dwayne Durdle foolishly and remorselessly pick on a little black kid named Tyrone Morris in the serving line in the cafeteria, and when Tyrone could take no more, he turned with a look of weariness and sad exasperation and threw a flurry of punches into Durdle's absorbent face so fast that you didn't actually see his hands move. All you heard was a kind of rubbery *flubba-da-dubba* sound and the *ping* of teeth ricocheting off walls and radiators. As Durdle sank to his knees, glassy-eyed and gurgling, Tyrone thrust an arm far down his throat, grabbed hold of something deep inside and turned him inside out.

'Goddam *fool* muthah-fuckah,' Tyrone said in amazed dismay as he retrieved his tray and continued on to the dessert section.

There were, however, almost no overt bad feelings between blacks and whites at Callanan. The black kids were poorer than the rest of us almost without exception, but otherwise were just the same in nearly all respects. They came from solid, hardworking homes. They spoke with identical voices, shopped at the same stores, wore the same clothes, went to the same movies. We were all just kids. Apart from my grandmother asking for nigger babies at Bishop's, I don't remember hearing a single racist remark in the whole of my upbringing.

I wouldn't pretend that we didn't notice that black kids were black, but it was as close to not noticing as you could get. It was much the same with other ethnic groups. Some years ago when I came to apply a pseudonym to one of my boyhood friends, I chose the name Stephen Katz partly in honour of a Des Moines drugstore called Katz's, which was something of a local institution in my childhood, and partly because I wanted a short name that was easy to type. Never did it occur to me that the name was Semitic. I never thought of *anybody* in Des Moines as being Jewish. I don't believe anyone did. Even when they had names like Wasserstein and Liebowitz, it was always a surprise to learn they were Jewish. Des Moines wasn't a very ethnic place.

Anyway, Katz wasn't Jewish. He was Catholic. And it was at Callanan that I met him when he was recruited by Doug Willoughby to join in an organized takeover of the Audio-Visual Club – a cunning but unusual move and a lasting testament to Willoughby's genius. Club members were put in charge of maintaining and showing the school's enormous cache of educational films. Whenever a teacher wanted to show a movie – and some teachers did little else because it meant they didn't have to teach or even spend much time in the classroom – a member of the elite AV team would wheel a projector to the room in question, expertly thread and loop the film through half a dozen sprockets and show the desired educational offering.

Historically, the AV Club was the domain of the school's geekiest students, as you would expect, but Willoughby at once saw the advantages the club offered to normal people. For one thing, it provided a key to the only locked space in the building to which students had access and where we could almost certainly smoke once he had cracked the ventilation problem (which he quickly did). Further, it gave access to an enormous supply of movies, including all the sex education films made between roughly 1938 and 1958. Finally, and above all, it provided a legitimate excuse to be at large in the empty hallways of Callanan during class time. If challenged by a teacher while roaming through the shiny corridors (and what a delightful, relaxing, privileged place school corridors are when empty) you could simply say: 'I'm just going up to the AV room to do some essential maintenance on a Bell and Howell 1040-Z,' which was in fact more or less true. What you didn't say was that you would also be smoking half a pack of Chesterfields while there.

So at Willoughby's behest, fifteen of us joined the

club, and as our first order of business voted all the existing members out. Only Milton Milton was allowed to stay as a sort of token geek and because he gave us half a bottle of crème de menthe he'd stolen from his dad's liquor cabinet and because he threatened to report us to his parents, the principal, the school board and the county sheriff, whom he dubiously claimed as a close family friend, if we didn't allow him to remain in the club.

The AV room was tucked away in an obscure corner of the building, upstairs and at the back. It was like the school attic. It contained a large assortment of old stage props, costumes, scripts, yearbooks from the 1920s and 1930s and dusty shelves of old films – hygiene movies, newsreels, sex education films, marijuana-will-melt-your-brain films and much else. We spent many happy hours showing the sex education films on the walls.

Willoughby discovered a film splicing kit and spent hours editing the films for his own amusement, putting goose-stepping Nazis into movies about the Oregon Trail and so on. His finest moment was in a sex education film when the narrative line 'Johnny had just experienced his first nocturnal emission' was immediately followed by a shot of Naval Academy cadets throwing their hats in the air.

It was in the AV Club, as I say, that I met a transfer student from the Catholic school system named Stephen Katz. I have never come close to doing the real Stephen Katz justice on any of the occasions I have put him in my books – no mortal author could – and I'm afraid I won't now except to say that he is the most extraordinary human being I have ever met, and in many ways the best. In those days he was the chipperest, friendliest, most party-ready human being the earth had ever known when sober and even more so when drunk, which he was much of the time even at the age of fourteen. I have never known anyone so drawn to, so amiably at home with, intoxicants. It was evident from the first moment that he was an engaging danger.

Often Katz and Willoughby and I skipped school and spent long days trying to get into Willoughby's older brother Ronald's chest of drawers. Ronald had an enormous collection of men's magazines, which he kept securely locked in a large chest in his bedroom. Ronald was the oldest, smartest and by far best behaved of the Willoughby boys – he was an altar boy, Explorer Scout, member of the student council, hall monitor, permanent asshole – and more cunning than his three brothers put together. Not only was every drawer in the chest locked with ingenuity, but each drawer when opened had been given an impenetrable lid that seemed to offer no way in at all. On top of all this, much of the room, from the doorknob to certain of the floorboards, was lethally booby-trapped. Depending on what the intruder touched or tampered with, he might receive a bracing electric shock or come under multiple attacks from flying missiles, falling weights, swinging hammers, lunging mousetraps or generous effusions of homemade pepper spray.

I particularly remember a moment of brief-lived delight when Willoughby, after hours of forensic examination, finally figured out how to open the second drawer of the chest – it had something to do with rotating a piece of carved filigree on the chest moulding – and in the same instant there came a whistling sound, and a slender homemade dart, about six inches long and beautifully made, embedded itself with a resonant *thwoing* in the chest not two inches to the left of Willoughby's fortuitously inclined head. Attached to the shaft of the dart was a slip of paper on which was neatly written: 'WARNING: I SHOOT TO KILL.'

'He's fucking crazy,' we agreed in unison.

After that Willoughby shrouded himself in every defensive item of apparel he could think of – welder's goggles, hockey mitts, heavy overcoat, catcher's chest protector, motorcycle helmet, and whatever else came to hand – while Katz and I hovered in the hallway urging him on and asking for updates on progress.

There was a particular urgency to the task because *Playboy* had lately taken to showing pubic hair. It is hard to believe that until the 1960s such an important erogenous zone remained undiscovered, but it is so. Prior to this, women in men's magazines had no reproductive apparatus at all – at least none that they were prepared to show to strangers. They seemed to suffer from an odd reflex medical condition – *vaginis timiditus*, Willoughby called it – that for some reason compelled them, whenever a camera was produced, to wrench their hips and fling one leg over another as if trying to get their lower half to face backwards. For years I thought that was the position women naturally adopted when they were naked and at ease. When *Playboy* first showed pubic hair, for at least seventy-two hours it featured in every male conversation in America. ('Check your oil for you, mister? Seen the new *Playboy* yet?') Woolworth's sold out its entire stock of magnifying glasses in twenty-four hours.

We longed with all our hearts to enter that privileged inner circle, as it were. But in over two years of trying, Willoughby never did get into his brother's private stock, until one day in frustration he broke open the bottom drawer with a fireman's axe, and a cornucopia of men's magazines – my goodness, but his brother was a collector – came sliding out. I have seldom passed a more agreeable or instructive afternoon. Willoughby was grounded for two months for that, but we all agreed it was a noble sacrifice, and he did have the satisfaction of getting his brother in trouble too, for some of these magazines were frankly quite disturbing.

As always, my timing with regard to actual female flesh remained impeccably abysmal. In the summer between eighth and ninth grades, I went away to visit my grandparents, where I had the usual delightful interludes with my Uncle Dee, the human flocking machine, and came back to find that in my absence a girl of radiant prettiness and good cheer named Kathy Wilcox had come to Willoughby's house to borrow some tracing paper and ended up teaching him and Katz a new game she had learned at Bible camp – at Bible camp!!!!! – in which you blindfolded a volunteer, spun the volunteer round for a couple of minutes, and then pressed firmly on his or her chest thirty or so times, at which point the victim would amusingly faint.

'Happens every time,' they said.

'I'm sorry, did you just say "her chest" – "pressed on *her* chest"?' I said.

Kathy Wilcox was a young woman with a chest worth pressing. The mere mention of her name was enough to make every corpuscle of blood in my body rush to the pelvic region and swell up in huge pointless readiness. They nodded happily. I couldn't believe this was happening to me again.

'Kathy Wilcox's chest? You were pressing on Kathy Wilcox's chest? With your hands?'

'Repeatedly,' said Willoughby, beaming.

Katz confirmed it with many happy nods.

My despair cannot be described. I had missed out on the only genuinely erotic, hands-on experience that there would ever be involving boys aged fourteen and instead had passed forty-eight hours watching a man turn assorted foods into flying whey.

Smoking was the big discovery of the age. Boy, did I love smoking and boy did it love me. For a dozen or so years I

347

did little in life but sit at desks hunched over books French inhaling (which is to say drawing ropes of smoke up into the nostrils from the mouth, which gives a double hit of nicotine with every heady inhalation as well as projecting an air of cerebral savoir faire, even at the cost of having a nicotine-stained upper lip and permanent yellowy-brown circlets about the nostrils) or lounge back with hands behind head blowing languorous smoke rings, at which I grew so proficient that I could bounce them off pictures on distant walls or fire one smoke ring through another – skills that marked me out as a Grand Master of smoking before I was quite fifteen.

We used to smoke in Willoughby's bedroom, sitting beside a window fan that was set up to blow outwards, so that all the smoke was pulled into the whirring blades and dispatched into the open air beyond. There was a prevailing theory in those days (of which my father was a devoted, and eventually solitary, advocate) that if the fan blew outwards it drew all the hot air from the room and pulled cool air in through any other open window. It was somehow supposed to be much more economical, which is where the appeal lay for my father. In fact, it didn't work at all – all it did was make the outside a little cooler – and pretty soon everyone abandoned it, except my father who continued to cool the air outside his window till his dying day.

Anyway, the one benefit of having a fan blow outwards was that it allowed you to finish each smoke with a flourish: you flicked the butt into the humming blades, which diced it into a shower of outward-flying sparks that was rather pleasing to behold and neatly obliterated the cigarette in the process, leaving no visible evidence below. It all worked very well until one August evening when Willoughby and I had a smoke, then went out for air, unaware that a solitary wayward ember had been flung back into the room and lodged in a fold of curtain, where it smouldered for an hour or so and then burst into a low but cheerful flame. When we returned to Willoughby's house, there were three fire trucks out front; fire hoses were snaked across the lawn, through the front door and up the stairs; Willoughby's bedroom curtains and several pieces of furniture were on the front lawn soaked through and still smoking lightly; and Mr Willoughby was on the front porch in a state of high emotion waiting to interview his son.

Mr Willoughby's troubles did not end there, however. The following spring, to celebrate the last day of the school year, Willoughby and his brother Joseph decided to make a bomb that they would pack in confetti and bury the night before in the centre of the Callanan lawn, a handsome sward of never-walked-upon grass enclosed by a formal semi-circular driveway. At 3.01 p.m., just as a thousand chattering students were pouring from the school's four exits, the bomb, activated by an alarm clock timer, would go off with an enormous bang that would fill the air with dirt and drifting smoke and a pleasing shower of twirling coloured paper.

The Willoughby brothers spent weeks mixing up dangerous batches of gunpowder in their bedroom and testing various concoctions, each more robust than the previous one, in the woods down by the railroad tracks near Waterworks Park. The last one left a smoking crater almost four feet across, threw strips of confetti twenty-five feet into the air and made such a reverberating, city-wide bang that squad cars hastened to the scene from eight different directions and cruised slowly around the area in a suspicious, squinty-eyed manner for almost forty minutes (making it the longest spell that Des Moines cops had ever been known to go without doughnuts and coffee).

It promised to be a fantastic show – the most memorable letting-out day in the history of Des Moines schools. The plan was that Willoughby and his brother would rise at four, walk to the school grounds under cover of darkness, plant the bomb and withdraw to await the end of the school day. To that end they assembled the necessary materials – spade, dark clothes, ski masks – and carefully prepared the bomb, which they left ticking away on the bedroom desk. Why they set the timer is a question that would be asked many times in the coming days. Each brother would vigorously blame the other. What is certain is that they retired to bed without its occurring to either of them that 3.01 *a.m.* comes before 3.01 *p.m.*

So it was at that dark hour, fifty-nine minutes before

their own alarm went off, that the peaceful night was rent by an enormous explosion in Doug and Joseph Willoughby's bedroom. No one in Des Moines was out at that hour, of course, but anyone passing who chanced to glance up at the Willoughbys' house at the moment of detonation would have seen first an intense yellow light upstairs, followed an instant later by two bedroom windows blowing spectacularly outwards, followed a second after that by a large puff of smoke and a cheery flutter of confetti.

But of course the truly memorable feature of the event was the bang, which was almost unimaginably robust and startling. It knocked people out of bed up to fourteen blocks away. Automatic alarms sounded all over the city, and the ceiling sprinklers came on in at least two office buildings. A community air raid siren was briefly activated, though whether by accident or as a precaution was never established. Within moments two hundred thousand groggy, bed-flung people were peering out their bedroom windows in the direction of one extremely well-lit, smoke-filled house on the west side of town, through which Mr Willoughby, confused, wild of hair, at the end of an extremely stretched tether, was stumbling, shouting: 'What the fuck? What the fuck?'

Doug and his brother, though comically soot-blackened and unable to hear anything not shouted directly into their ears for the next forty-eight hours, were miraculously unharmed. The only casualty was a small laboratory rat that lived in a cage on the desktop and was now just a lot of disassociated fur. The blast knocked the Willoughby home half an inch off its foundations and generated tens of thousands of dollars in repair bills. The police, fire department, sheriff's office and FBI all took a keen interest in prosecuting the family, though no one could ever quite agree on what charges to bring. Mr Willoughby became involved in protracted litigation with his insurers and embarked on a long programme of psychotherapy. In the end, the whole family was let off with a warning. Doug Willoughby and his brother were not allowed off the property except to go to school or attend confession for the next six months. Technically, they are still grounded.

And so we proceeded to high school.

Drinking became the preoccupation of these tall and festively pimpled years. All drinking was led by Katz, for whom alcohol was not so much a pastime as a kind of oxygen. It was a golden age for misbehaviour. You could buy a six-pack of Old Milwaukee beer for 59 cents (69 cents if chilled) and a pack of cigarettes (Old Gold was the brand of choice for students of my high school, Roosevelt, for no logical or historic reason that I am aware of) for 35 cents, and so have a full evening of pleasure for less than a dollar, even after taking into account sales tax. Unfortunately it was impossible to buy beer, and nearly as difficult to buy cigarettes, if you were a minor.

Katz solved this problem by becoming Des Moines's most accomplished beer thief. His career of crime began in seventh grade when he hit on a scheme that was simplicity itself. Dahl's, as part of its endless innovative efficiency, had coolers that opened from the back as well as the front so that they could be stocked from behind from the storeroom. Also inside the storeroom was a wooden pen filled with empty cardboard boxes waiting to be flattened and taken away for disposal. Katz's trick was to approach a member of staff by the stock-room door and say, 'Excuse me, mister. My sister's moving to a new apartment. Can I take some empty boxes?'

'Sure, kid,' the person would always say. 'Help yourself.'

So Katz would go into the stock room, select a big box, load it quickly with delicious frosty beer from the neighbouring beer cooler, put a couple of other boxes on top as cover, and stroll out with a case of free beer. Often the same employee would hold the door open for him. The hardest part, Katz once told me, was acting as if the boxes were empty and didn't weigh anything at all.

Of course you could ask for boxes on only so many occasions without raising suspicion, but fortunately there were Dahl's stores all over Des Moines with the same help-yourself coolers, so it was just a matter of moving around from store to store. Katz got away with it for over two years and would be getting away with it still, I daresay, except that the bottom gave way on a box once at the

Dahl's in Beaverdale as Katz was egressing the building, and sixteen quart bottles of Falstaff smashed on to the floor in a foamy mess. Katz was not built for running, and so he just stood grinning until a member of staff strolled over and took him unresisting to the manager's office. He spent two weeks at Meyer Hall, the local juvenile detention centre, for that.

I had nothing to do with store thefts. I was far too cowardly and prudent to so conspicuously break the law. My contribution was to forge drivers' licences. These were, if I say it myself, small masterpieces – albeit bearing in mind that state drivers' licences were not terribly sophisticated in those days. They were really just pieces of heavy blue paper, the size of a credit card, with a kind of wavy watermark. My stroke of brilliance was to realize that the back of my father's cheques had almost exactly the same wavy pattern. If you cut one of his cheques to the right size, turned it over and, with the aid of a T-square, covered the blank side with appropriate-sized boxes for the bearer's name and address and so on, then carefully inked the words 'Iowa Department of Motor Vehicles' across the top with a fine pen and a straight edge, and produced a few other small flourishes, you had a pretty serviceable fake driver's licence.

If you then put the thing through an upright office typewriter such as my father's, entering false details in the little boxes, and in particular giving the bearer a suitably early date of birth, you had a product that could be taken to any small grocery store in town and used to acquire limitless quantities of beer.

What I didn't think of until much too late was that the obverse side of these homemade licences sometimes bore selected details of my father's account – bank name, account number, tell-tale computer coding and so on – depending on which part of the cheque I had cut to size with scissors.

The first time this occurred to me was about 9.30 a.m. on a weekday when I was summoned to the office of the Roosevelt principal. I had never visited the principal's office before. Katz was there already, in the outer waiting room. He was often there.

'What's up?' I said.

But before he could speak I was called into the inner sanctum. The principal was sitting with a plainclothes detective who introduced himself as Sergeant Rotisserie or something like that. He had the last flat-top in America.

'We've uncovered a ring of counterfeit driver's licences,' the sergeant told me gravely and held up one of my creations.

'A *ring*?' I said and tried not to beam. My very first foray into crime and already I was, single-handedly, a 'ring'. I couldn't have been more proud. On the other hand, I didn't particularly want to be sent to the state reform school at Clarinda and spend the next three years having involuntary soapy sex in the showers with guys named Billy Bob and Cletus Leroy.

He passed the licence to me to examine. It was one I had done for Katz (or 'Mr B. Bopp', as he had rather rakishly restyled himself). He had been picked up while having a beer-induced nap on the grassy central reservation of Polk Boulevard the night before and a search of his personal effects at the station house had turned up the artificial licence, which I examined with polite interest now. On the back it said 'Banker's Trust' and beneath that was my father's name and address – something of a giveaway to be sure.

'That's your father, isn't it?' said the detective.

'Why, yes it is,' I answered and gave what I hoped was a very nice frown of mystification.

'Like to tell me how that happened?'

'I can't imagine,' I said, looking earnest, and then added: 'Oh, wait. I bet I know. I had some friends over last week to listen to records, you know, and some fellows we'd never seen before crashed in on us, even though it wasn't even a party.' I lowered my voice slightly. 'They'd been drinking.'

The detective nodded grimly, knowledgeably. He'd been to this slippery slope before.

'We asked them to leave, of course, and eventually they did when they realized we didn't have any beer or other intoxicants, but I just bet you while we weren't looking one of them went through my dad's desk and stole some cheques.'

'Any idea who they were?'

'I'm pretty sure they were from North High. One of them looked like Richard Speck.'

The detective nodded. 'It starts to make sense, doesn't it? Do you have any witnesses?'

'Oom,' I said, a touch noncommittally, but nodded as if it might be many.

'Was Stephen Katz present?'

'I think so. Yes, I believe he was.'

'Would you go out and wait in the outer room and tell Mr Katz to come in?'

I went out and Katz was sitting there. I leaned over to him and said quickly: 'North High. Crashed party. Stole cheques. Richard Speck.'

He nodded, instantly understanding. This is one of the reasons why I say Katz is the finest human being in the world. Ten minutes later I was called back in.

'Mr Katz here has corroborated your story. It appears these boys from North High stole the cheques and ran them through a printing press. Mr Katz here was one of their customers.'

He looked at Katz without much sympathy.

'Great! Case solved!' I said brightly. 'So, can we go?'

'You can go,' said the sergeant. 'I'm afraid Mr Katz will be coming downtown with me.'

So Katz took the rap, allowing me to keep a clean sheet, God bless him and keep him. He spent a month in juvenile detention.

* * *

The thing about Katz was that he didn't do bad things with alcohol because he wanted to; he did them because he needed to. Casting around for a new source of supply, he set his sights higher. Des Moines had four beer distribution companies, all in brick depots in a quiet quarter at the edge of downtown where the railroad tracks ran through. Katz watched these depots closely for a couple of weeks and realized that they had practically no security and never worked on Saturdays or Sundays. He also noticed that railroad boxcars often stood in sidings beside the depots, particularly at weekends.

So one Sunday morning Katz and a kid named Jake Bekins drove downtown, parked beside a boxcar and knocked off its padlock with a sledgehammer. They slid open the boxcar door and discovered that it was filled solid with cases of beer. Wordlessly they filled Bekins's car with boxes of beer, shut the boxcar door and drove to the house of a third party, Art Froelich, whose parents were known to be out of town at a funeral. There, with Froelich's help, they carried the beer down to the basement. Then the three of them went back to the boxcar and repeated the process. They spent the whole of Sunday transferring beer from the boxcar to Froelich's basement until they had emptied the one and filled the other.

Froelich's parents were due home on Tuesday, so on Monday Katz and Bekins got twenty-five friends to put up $5 each and they rented a furnished apartment in an easygoing area of town known as Dogtown near Drake University. Then they transferred all the beer by car from Froelich's basement to the new apartment. There Katz and Bekins drank seven evenings a week and the rest of us dropped in for a Schlitz cocktail after school and for more prolonged sessions at weekends.

Three months later all the beer was gone and Katz and a small corps of henchmen returned downtown and spent another Sunday emptying out another boxcar from another distributor. When, three months after that, they ran out of beer again, they ventured downtown once more, but more cautiously this time because they were certain that after two big robberies somebody would be keeping a closer watch on the beer warehouses.

Remarkably, this seemed not to be so. This time there were no boxcars, so they knocked a panel out of one of the warehouse delivery-bay doors, and slipped through the hole. Inside was more beer than they had ever seen at once – stacks and stacks of it standing on pallets and ready to be delivered to bars and stores all over central Iowa on Monday.

Working nonstop, and drafting in many willing assistants, they spent the weekend loading cars one after another with beer and slowly emptying the warehouse. Froelich expertly worked a forklift and Katz directed traffic. For a whole miraculous weekend, a couple of dozen high school kids could be seen – if anyone had bothered to look – moving loads of beer out of the warehouse, driving it across town and carrying it in relays into a slightly sagging and decrepit apartment house on Twenty-third Street and Forest Avenue. As word got around, other kids from other high schools began turning up, asking if they could take a couple of cases.

'Sure,' Katz said generously. 'There's plenty for all. Just pull your car up over there and try not to leave any finger-prints.'

It was the biggest heist in Des Moines in years, possibly ever. Unfortunately so many people became involved that everyone in town under the age of twenty knew who was responsible for it. No one knows who tipped off the police, but they arrested twelve principal conspirators in a dawn raid three days after the theft and took them all downtown in handcuffs for questioning. Katz was of course among them.

These were good kids from good homes. Their parents were mortified that their offspring could be so wilfully unlawful. They called in expensive lawyers, who swiftly cut deals with the prosecutor to drop charges if they named names. Only Katz's parents wouldn't come to an arrangement. They couldn't comfortably afford to and anyway they didn't believe it was right. Besides, *somebody* had to take the rap – you can't just let every guilty person go or what kind of criminal justice system would you have, for goodness' sake? – so it was necessary to elect a fall guy and everyone agreed that Katz should be that person. He was charged with grand theft, a felony, and sent to reform school for two years. It was the last we saw of him till college.

I got through high school by the skin of my teeth. It was my slightly proud boast that I led the school in absences all three years and in my junior year achieved the distinction of missing more days than a boy with a fatal illness, as Mrs Smolting, my careers counsellor, never tired of reminding me. Mrs Smolting hated me with a loathing that was slightly beyond bottomless.

'Well, frankly, William,' she said with a look of undisguised disdain one day after we had worked our way through a long list of possible careers, including vacuum cleaner repair and selling things door to door, and established to her absolute satisfaction that I lacked the moral fibre, academic credentials, intellectual rigour and basic grooming skills for any of them, 'it doesn't appear that you are qualified to do much of anything.'

'I guess I'll have to be a high school careers counsellor then!' I quipped lightly, but I'm afraid Mrs Smolting did not take it well. She marched me to the principal's office – my second visit in a season! – and lodged a formal complaint.

I had to write a letter of abject apology, expressing respect for Mrs Smolting and her skilled and caring profession, before they would allow me to continue to my senior year, which was a serious business indeed because at this time, 1968, the only thing that stood between one's soft tissue and a Vietcong bullet was the American education system and its automatic deferment from the draft. A quarter of young American males were in the armed forces in 1968. Nearly all the rest were in school, in prison or were George W. Bush. For most people, school was the only realistic option for avoiding military service.

In one of his last official acts, but also one of his most acclaimed ones, the Thunderbolt Kid turned Mrs Smolting into a small hard carbonized lump of a type known to people in the coal-burning industry as a clinker. Then he handed in a letter of carefully phrased apology, engaged in a few months of light buckling down, and graduated, unshowily, near the bottom of his class.

The following autumn he enrolled at Drake, the local university. But after a year or so of desultory performance there, he went to Europe, settled in England and was scarcely ever heard from again.

Chapter 14
FAREWELL

In Milwaukee, uninjured when his auto swerved off the highway, Eugene Cromwell stepped out to survey the damage and fell into a 50-foot limestone quarry. He suffered a broken arm.

– *Time* magazine, 23 April 1956

From time to time when I was growing up, my father would call us into the living room to ask how we felt about moving to St Louis or San Francisco or some other big-league city. The *Chronicle* or *Examiner* or *Post-Dispatch*, he would inform us sombrely, had just lost its baseball writer – he always made it sound as if the person had not returned from a mission, like a Second World War airman – and the position was being offered to him.

'Money's pretty good, too,' he would say with a look of frank consternation, as if surprised that one could be paid for routinely attending Major League baseball games.

I was always for it. When I was small, I was taken with the idea of having a dad working in a field where people evidently went missing from time to time. Then later it was more a desire to pass what remained of my youth in a place – any place at all – where daily hog prices were not regarded as breaking news and corn yields were never mentioned.

But it never happened. In the end he and my mother always decided that they were content in Des Moines. They had good jobs at the *Register* and a better house than we could afford in a big city like San Francisco. Our friends were there. We were settled. Des Moines felt like, Des Moines was, home.

Now that I am older I am glad we didn't leave. I have a lifelong attachment to the place myself, after all. Every bit of formal education I have ever had, every formative experience, every inch of vertical growth on my body took place within this wholesome, friendly, nurturing community.

Of course much of the Des Moines I grew up in is no longer there. It was already changing by the time I reached adolescence. The old downtown movie palaces were among the first to go. The Des Moines Theater, that wonderful heap of splendour, was torn down in 1966 to make way for an office building. I didn't realize it until I read a history of the city for this book, but the Des Moines was not just the finest theatre in the city, but possibly the finest surviving theatre of any type between Chicago and the West Coast. I was further delighted to discover that it had been built by none other than A. H. Blank, the philanthropist with the penthouse apartment that Jed Mattes and I used to visit. He had spent the exceptionally lavish sum of $750,000 on the building in 1918. It is extraordinary to think that it didn't even survive for half a century. The other principal theatres of my childhood – the Paramount, Orpheum (later called the Galaxy), Ingersoll, Hiland, Holiday and Capri – followed one by one. Nowadays if you want to see a movie you have to drive out to a shopping mall, where you can choose between a dozen pictures, but just one very small size of screen, each inhabiting a kind of cinematic shoebox. Not much magic in that.

Riverview Park closed in 1978. Today it's just a large vacant lot with nothing to show that it ever existed. Bishop's, our beloved cafeteria, closed about the same time, taking its atomic toilets, its little table lights, its glorious foods and kindly waitresses with it. Many other locally owned restaurants – Johnny and Kay's, Country Gentleman, Babe's, Bolton and Hay's, Vic's Tally-Ho, the beloved Toddle House – went around the same time. Katz helped the Toddle House on its way by introducing to it a new concept called 'dine and dash', in which he and whoever he had been drinking with would consume a hearty late-night supper and then make a hasty exit without paying, calling over their shoulder if challenged, 'Short of cash – gotta dash!' I wouldn't say that Katz single-handedly put the Toddle House out of business, but he didn't help.

The *Tribune*, the evening paper which I lugged thanklessly from house to house for so many years, closed in 1982 after it was realized that no one had actually been reading it since about 1938. The *Register*, its big sister, which once truly was the pride of Iowa, got taken over by the Gannett organization three years later. Today it is, well, not what it was. It no longer sends a reporter to baseball spring training or even always to the World Series, so it is perhaps as well my father is no longer around.

Greenwood, my old elementary school, still

commands its handsome lawn, still looks splendid from the street, but they tore out the wonderful old gym and auditorium, its two most cherishable features, to make way for a glassy new extension out back, and the other distinguishing touches – the cloakrooms, the clanking radiators, the elegant water fountains, the smell of mimeograph – are long gone, too, so it's no longer really the place I knew.

My peerless Little League park, with its grandstand and press box, was torn down so that somebody could build an enormous apartment building in its place. A new, cheaper park was built down by the river bottoms near where the Butters used to live, but the last time I went down there it was overgrown and appeared to be abandoned. There was no one to ask what happened because there are no people outdoors any more – no kids on bikes, no neighbours talking over fences, no old men sitting on porches. Everyone is indoors.

Dahl's supermarket is still there, and still held in some affection, but it lost the Kiddie Corral and grocery tunnel years ago during one of its periodic, and generally dismaying, renovations. Nearly all the other neighbourhood stores – Grund's Groceries, Barbara's Bake Shoppe, Reed's ice cream parlour, Pope's barbershop, the Sherwin-Williams paint store, Mitcham's TV and Electrical, the little shoe repair shop (run by Jimmy the Italian – a beloved local figure), Henry's Hamburgers, Reppert's Drugstore – are long gone. Where several of them stood there is now a big Walgreen's drugstore, so you can buy everything under one roof in a large, anonymous, brightly lit space from people who have never seen you before and wouldn't remember you if they had. It stocks men's magazines, I was glad to note on my last visit, though these are sealed in plastic bags, so it is actually harder now to see pictures of naked women than it was in my day, which I would never have believed possible, but there you are.

All the downtown stores went one by one. Ginsberg's and the New Utica department stores closed. Kresge's and Woolworth's closed. Frankel's closed. Pinkie's closed. J. C. Penney bravely opened a new downtown store and that closed. The Shops Building lost its restaurant. Then somebody got mugged or saw a disturbed homeless person or something, and nobody went downtown after dark at all ever, and all the rest of the restaurants and night spots closed. In the ultimate indignity, even the bus station moved out.

Younkers, the great ocean liner of a department store, became practically the last surviving relic of the glory days of my childhood. For years it heroically held on in its old brown building downtown, though it closed whole floors and retreated into ever tinier corners of the building to survive. In the end it had only sixty employees, compared with over a thousand in its heyday. In the summer of 2005, after one hundred and thirty-one years in business, it closed for the last time.

* * *

When I was a kid, the Register and Tribune had an enormous photo library, in a room perhaps eighty feet by sixty feet, where I would often pass an agreeable half-hour if I had to wait for my mom. There must have been half a million pictures in there, maybe more. You could look in any drawer of any filing cabinet and find real interest and excitement from the city's past – five-alarm fires, train derailments, a lady balancing beer glasses on her bosom, parents standing on ladders at hospital windows talking to their polio-stricken children. The library was the complete visual history of Des Moines in the twentieth century.

Recently I returned to the R & T looking for illustrations for this book, and discovered to my astonishment that the picture library today occupies a small room at the back of the building and that nearly all the old pictures were thrown out some years ago.

'They needed the space,' Jo Ann Donaldson, the present librarian, told me with a slightly apologetic look.

I found this a little hard to take in. 'They didn't give them to the state historical society?' I asked.

She shook her head.

'Or the city library? Or a university?'

She shook her head twice more. 'They were recycled for the silver in the paper,' she told me.

So now not only are the places mostly gone, but there is no record of them either.

* * *

Life moved on for people, too – or in some unfortunate cases stopped altogether. My father slipped quietly into the latter category in 1986 when he went to bed one night and didn't wake up, which is a pretty good way to go if you have to go. He was just shy of his seventy-first birthday when he died. Had he worked for a bigger newspaper, I have no doubt my father would have been one of the great baseball writers of his day. Because we stayed, the world never got a chance to see what he could do. Nor, of course, did he. In both cases, I can't help feeling that they didn't know what they were missing.

My mother stayed on in the family home for as long as she could manage, but eventually sold up and moved to a nice old apartment building on Grand. Now in her nineties, she remains gloriously cheerful, healthy and perky, keen as ever to spring up and make a sandwich from some Tupperwared memento at the back of her fridge. She still keeps an enormous stock of jars under the sink (though none has ever experienced a drop of toity, she assures me) and retains one of the Midwest's most outstanding collections of sugar packets, saltine crackers and jams of many flavours. She would like the record to show, incidentally, that she is nothing like as bad a cook as her feckless son persists in portraying her in his books, and I am happy to state here that she is absolutely right.

As for the others who passed through my early life

and into the pages of this book, it is difficult to say too much without compromising their anonymity.

Doug Willoughby had what might be called a lively four years at college – it was an age of excess; I'll say no more – but afterwards settled down. He now lives quietly and respectably in a small Midwestern city, where he is a good and loving father and husband, a helpful neighbour and supremely nice human being. It has been many years since he has blown anything up.

Stephen Katz left high school and dived head first into a world of drugs and alcohol. He spent a year or two at the University of Iowa, then returned to Des Moines, where he lived near the Timber Tap, a bar on Forest Avenue which had the distinction of opening for business at 6 a.m. every day. Katz was often to be seen at that hour entering in carpet slippers and a robe for his morning 'eye-opener'. For twenty-five years or so, he took into his body pretty much whatever consciousness-altering replenishments were on offer. For a time he was one of only two opium addicts in Iowa (the other was his supplier) and famous among his friends for a remarkable ability to crash cars spectacularly and step from the wreckage grinning and unscathed. After taking a leading role in a travel adventure story called A Walk in the Woods (which he describes as 'mostly fiction'), he became a respectful and generally obedient member of Alcoholics Anonymous, landed a job in a printing plant, and found a saintly life partner named Mary. At the time of writing, he had just passed his third-year anniversary of complete sobriety – a proud achievement.

Jed Mattes, my gay friend, moved with his family to Dubuque soon after he treated me to the strippers' tent at the State Fair, and I lost touch with him altogether. Some twenty years later when I was looking for a literary agent, I asked a publishing friend in New York for a recommendation. He mentioned a bright young man who had just quit the ICM literary agency to set up on his own. 'His name's Jed Mattes,' he told me. 'You know, I think he might be from your hometown.'

So Jed became my agent and close renewed friend for the next decade and a half. In 2003, after a long battle with cancer, he died. I miss him a great deal. Jed Mattes is, incidentally, his real name – the only one of my contemporaries, I believe, to whom I have not given a pseudonym.

Buddy Doberman vanished without trace halfway through college. He went to California in pursuit of a girl and was never seen again. Likewise of unknown fate were the Kowalski brothers, Lanny and Lumpy. Arthur Bergen became an enormously rich lawyer in Washington, DC. The Butter clan went away one springtime and never returned. Milton Milton went into the military, became something fairly senior, and died in a helicopter crash during the preparations for the first Gulf War.

Thanks to what I do, I sometimes renew contact with people unexpectedly. A woman came up to me after a reading in Denver once and introduced herself as the former Mary O'Leary. She had on big glasses that she kept round her neck on a chain and seemed jolly and happy and quite startlingly meaty. On the other hand, a person I had thought of as timid and mousy came up to me at another reading and looked like a movie star. I think life is rather splendid like that.

The Thunderbolt Kid grew up and moved on. Until quite recently he still occasionally vaporized people, usually just after they had walked through a held door without saying thank you, but eventually he stopped eliminating people when he realized that he couldn't tell which of them buy books.

The Sacred Jersey of Zap, moth-eaten and full of holes, was thrown out in about 1978 by his parents during a tragically misguided housecleaning exercise, along with his baseball cards, comic books, *Boys' Life* magazines, Zorro whip and sword, Sky King neckerchief and neckerchief ring, Davy Crockett coonskin cap, Roy Rogers decorative cowboy vest and bejewelled boots with jingly tin spurs, official Boy Scout Vitt-L-Kit, Sky King Fan Club card and other related credentials, Batman flashlight with signalling attachment, electric football game, Johnny Unitas-approved helmet, Hardy Boys books and peerless set of movie posters, many in mint condition.

That's the way of the world, of course. Possessions get discarded. Life moves on. But I often think what a shame it is that we didn't keep the things that made us different and special and attractive in the Fifties. Imagine those palatial downtown movie theatres with their vast screens and Egyptian decor, but thrillingly enlivened with Dolby sound and slick computer graphics. Now that *would* be magic. Imagine having all of public life – offices, stores, restaurants, entertainments – conveniently clustered in the heart of the city and experiencing fresh air and daylight each time you moved from one to another. Imagine having a cafeteria with atomic toilets, a celebrated tearoom that gave away gifts to young customers, a clothing store with a grand staircase and a mezzanine, a Kiddie Corral where you could read comics to your heart's content. Imagine having a city full of things that no other city had.

What a wonderful world that would be. What a wonderful world it was. We won't see its like again, I'm afraid.

BIBLIOGRAPHY

The following are books mentioned or alluded to in the text:

Castleman, Harry, and Walter J. Podrazik, *Watching TV: Six Decades of American Television*. Syracuse, New York: Syracuse University Press, 2003.

DeGroot, Gerard J., *The Bomb: A Life*. Cambridge, Massachusetts: Harvard University Press, 2005.

Denton, Sally, and Roger Morris, *The Money and the Power: The Making of Las Vegas and Its Hold on America, 1947–2000*. London: Pimlico, 2002.

Diggins, John Patrick, *The Proud Decades: America in War and Peace, 1941–1960*. New York: W. W. Norton, 1988.

Goodchild, Peter, *Edward Teller: The Real Dr Strangelove*. London: Weidenfeld and Nicolson, 2004.

Halberstam, David, *The Fifties*. New York: Fawcett Columbine, 1993.

Heimann, Jim (ed.), *The Golden Age of Advertising – the 50s*. Cologne: Taschen, 2002.

Henriksen, Margot A., *Dr Strangelove's America: Society and Culture in the Atomic Age*. Berkeley: University of California Press, 1997.

Kismaric, Carole, and Marvin Heiferman, *Growing Up with Dick and Jane: Learning and Living the American Dream*. San Francisco: Lookout/HarperCollins, 1996.

Lewis, Peter, *The Fifties*. London: Heinemann, 1978.

Light, Michael, *100 Suns: 1945–1962*. London: Jonathan Cape, 2003.

Lingeman, Richard R., *Don't You Know There's a War On?: The American Home Front 1941–1945*. New York: G. P. Putnam's Sons, 1970.

McCurdy, Howard E., *Space and the American Imagination*. Washington: Smithsonian Institution Press, 1997.

Mills, George, *Looking in Windows: Surprising Stories of Old Des Moines*. Ames, Iowa: Iowa State University Press, 1991.

Oakley, J. Ronald, *God's Country: America in the Fifties*. New York: Dembner Books, 1986.

O'Reilly, Kenneth, *Hoover and the Un-Americans*. Philadelphia: Temple University Press, 1983.

Patterson, James T., *Grand Expectations: The United States, 1945–1974*. New York: Oxford University Press, 1996.

Savage, Jr., William W., *Comic Books and America, 1945–1954*. Norman: University of Oklahoma Press, 1990.

ILLUSTRATIONS

The Bryson family photos on pp. 7, 12, 14, 52, 130, 244, 364 and 381 are from the author's own collection.

pp. 4–5: An average American family and all the food they consumed in 1951. Hagley Museum and Library, Wilmington, Delaware

p. 16: Locust Street, Des Moines, 16 February 1953. *Dope Inferno* is showing at the cinema. State Historical Society of Iowa

p. 78: The Paramount movie palace, Des Moines, 1950s. *The Florodora Girl* is showing with Al Morey and Marion Davies. State Historical Society of Iowa

p. 104: Advertisement for Camel cigarettes, with a medical endorsement. Courtesy the Advertising Archives, London

p. 156: 'I dreamed I went to the office in my Maidenform bra'. Courtesy the Advertising Archives, London

p. 176: Naval observers watching a nuclear blast in the Pacific in the 1950s. © CORBIS

p. 198: Schoolchildren practising duck and cover drill in case of nuclear attack, February 1951. © Bettmann/CORBIS

p. 222: Mary McGuire, homecoming queen, in the Drake University Year Book of 1938. Special Collection at Cowles Library, Drake University, Des Moines

p. 270: Chesley Bonestell's vision of Manhattan under nuclear attack. *Collier's*, 5 August 1950. Courtesy Bonestell Space Art

p. 292: Spectators at the Iowa State Fair cake display, Des Moines, 1955. John Dominis/Timepix

p. 328: Charles van Doren on *Twenty-One*, the TV quiz show, 11 March 1957. It was later discovered that the show had been rigged. © Bettmann/CORBIS